Exploring Surveys and Information from Samples

Exploring Surveys and Information from Samples was prepared under the auspices of the American Statistical Association—National Council of Teachers of Mathematics Joint Committee on the Curriculum in Statistics and Probability.

This book is part of the Quantitative Literacy Project, which was funded in part by the National Science Foundation.

Exploring Surveys and Information from Samples

James M. Landwehr
AT&T Bell Laboratories
Murray Hill, New Jersey

Jim Swift
Nanaimo School District
Nanaimo, British Columbia

Ann E. Watkins
Los Angeles Pierce College
Woodland Hills, California

DALE SEYMOUR PUBLICATIONS

Acknowledgments

Grateful acknowledgment is made to the following:

The Gallup Organization, Inc. for the article on pages 59–60, "Design of the Gallup Sample." Reprinted by permission.

The New York Times for the article on page 58, "How the Poll Was Conducted," from the June 5, 1985 edition. Copyright © 1985 by the New York Times Company. Reprinted by permission.

Cover: John Edeen and Francesca Angelesco
Editing: Adrienne Harris
Technical Art: Pat Rogondino
Illustrations: John Johnson

This publication was prepared as part of the American Statistical Association Project—Quantitative Literacy—with partial support of the National Science Foundation Grant No. DPE-8317656. Any opinions, findings, conclusions, or recommendations expressed in this publication are those of the authors and do not necessarily represent the views of the National Science Foundation. These materials shall be subject to a royalty-free, irrevocable, worldwide, nonexclusive license in the United States Government to reproduce, perform, translate, and otherwise use and to authorize others to use such materials for Government purposes.

ISBN 0-86651-339-6

Order number DS01707

DALE SEYMOUR PUBLICATIONS
P.O. BOX 10888
PALO ALTO, CA 94303

5 6 7 8 9 10 11 12 13-MA-95 94 93 92

CONTENTS

PREFACE

Sample surveys provide useful and current information about our people, businesses, and governments. The growth in the use of sample surveys in the last 30 years has been unprecedented. Data from the Consumer Expenditure Survey, reflected in the Consumer Price Index, are used in collective bargaining agreements. Data collected in the Health Interview Survey measure the success of anti-smoking campaigns. The unemployment rate from the Current Population Survey is an important economic indicator making the news every month. In addition, survey data are now used to establish campaign issues for political candidates, to determine the survival of television programs, to set wage rates in certain industries, to locate new stores, and to determine the effectiveness of advertising.

Surveys are carried out by federal, state, and local governments, as well as by universities, businesses, private companies, political candidates, and non-profit groups. The results of these surveys appear in magazines and newspapers and are discussed extensively on television and radio. Because of the growth in the survey industry, there are many different kinds of people doing surveys. Some of these groups have people well-trained in statistical methods guiding their effort; other groups do not realize that there is a statistical basis to sample surveys.

All survey organizations depend on the public for two important reasons. The public provides survey data. Well-designed surveys suffer when people refuse to be interviewed. Some people refuse because surveys seem unimportant or because they don't understand how the views of a sample can represent the views of an entire population. If you, as students, gain a greater understanding of how surveys work and what makes them useful, the value of surveys that you may be asked to participate in will improve.

The public also contains data users. Many people use statistics generated from sample surveys—to see where their candidates stand in polls, to evaluate how well government programs are working, to find out where a new store should be located. Users must question the accuracy of survey results and ask the right questions. Was the sample size large enough so that you can have any confidence in the statistics produced? What were the questions asked? Were there any problems in carrying out the survey that you should know about? To the extent that this book helps you understand the strengths and weaknesses of survey results, you will be able to recognize and use good survey data more effectively.

As an Associate Director of the Bureau of the Census, I applaud the arrival of this book in the classroom. Teaching young people to understand and use sample surveys today will surely result in better surveys and better uses of survey data in the future.

Barbara A. Bailar
Associate Director for Statistical Standards
   and Methodology, United States Bureau
   of the Census
President, American Statistical Association, 1987

I. INTRODUCTION

The United States Constitution requires "enumeration" of the population in order to determine how many seats each state should have in the 435-member House of Representatives. Thus, every 10 years, the Bureau of the Census attempts to count the entire population of the United States. Taking a **census** of the United States is incredibly expensive and difficult. The 1980 census of 86 million households required seven years of planning and about 280,000 workers. The questionnaires filled so many boxes that, if stacked up, they would have been 30 miles high.

The government also needs information about its citizens in the years between censuses. For example, to plan government programs, elected representatives must know how many people are unemployed, poor, and sick. Since the early 1940s, the government has used **sample surveys** to gather this information. Of the approximately 250 surveys taken by the Bureau of the Census each year, the best known is the Current Population Survey (CPS). This monthly survey estimates unemployment, income, schooling, and other measures by questioning about 100,000 people. Based on these people's responses, the bureau estimates the level of unemployment, for example, in the entire U.S. population. The unemployment figures you see on television or in the newspaper come from the CPS.

Another U.S. survey is the National Crime Survey, which the government began in the early 1970s to determine the extent of crime in the United States. Government workers cannot gather this information from police reports because the survey has revealed that people report only about 35% of all crimes to the police. For this survey, interviewers talk to people in about 60,000 households twice a year. (You can see a page from the National Crime Survey questionnaire used by interviewers on page 49.)

The type of survey reported most often in newspapers and on television is the **opinion poll**. The names of the leading polling organizations—Gallup, Roper, Harris, New York Times/CBS—are familiar to most adults. These organizations ask people about their political opinions, the consumer products they prefer, and their views on religion and education. People use the information for everything from planning a presidential candidate's campaign strategy to deciding the flavor of a new toothpaste.

This book will help you understand how statisticians can make statements about an entire group of people, or **population**, after they have questioned only a **sample** from that population. We will study only surveys (or polls) that ask questions people can answer with "yes" or "no." Here is an example of this type of survey.

The March 1985 Gallup survey asked 1,571 American adults this question:

"Do you approve or disapprove of the way Ronald Reagan is handling his job as president?"

Fifty-six percent said that they approved. For results based on samples of this size, one can say with 95% confidence that the error attributable to sampling and other random effects could be 3 percentage points in either direction.

In addition to sampling error, the reader should bear in mind that question wording and practical difficulties encountered in conducting surveys can introduce error or bias into the findings of opinion polls.

Source: Santa Barbara, California, *News-Press*, April 7, 1985.

Gallup surveyed (or polled) a sample of 1,571 adults from a total population of about 170 million adults. Pollsters asked each adult a *yes-no* question. ("Do you approve or disapprove . . . ?") The proportion of *yes* (or approve) responses from this sample was 0.56. The responses of the 1,571 adults might not exactly match those of the entire population. However, based on his calculations, Gallup feels confident that if he polled the entire American adult population, between 53% and 59% of the people would approve (a range of 3 percentage points in either direction from the 56%).

Gallup's statement that between 53% and 59% of the population would approve is a **statistical inference** he made about the population from the sample. In this book, you will learn the basic mathematics and statistics behind such an inference, and you will learn how to interpret the inference.

We will obtain samples from objects in containers, from coins, and from random number tables. During a one-hour class, you will find it much easier to take samples from a container than to take samples of the U.S. population!

Application 1

Guessing the Percentage of Yeses

Your teacher has a container of objects. Some of them are different from the rest; we will call them *yeses*. The ***population percentage*** is the percentage of objects in the container that are *yeses*. This percentage can be found exactly only by examining all the objects in the container. However, if we take a sample of objects from the container, we can estimate the population percentage by the ***sample proportion***. You can find the sample proportion by dividing the number of *yeses* in the sample by the sample size.

1. Mix up the objects and, without looking, take a sample of 10 objects from the container. What is the number of *yeses*?

2. Using the result from question 1, estimate the percentage of objects in the container that are *yeses*.

3. Give an interval around your estimate that is as small as possible but that you believe contains the population percentage. For example, if you get a sample proportion of 0.60, you may believe that the container has from 55% to 65% *yeses*.

4. What is the actual percentage of *yeses* in the container? (Your teacher will tell you.) Does your interval contain this percentage?

In this book, you will learn a method of constructing an interval, called a ***confidence interval***, that will contain the true percentage of *yeses* for most samples. We could, of course, let the interval be 0% to 100%, so that we are sure that the true percentage of *yeses* will be in the interval. But if Gallup, for example, reports that he is confident that between 0% and 100% of the population approve of the way the president is handling his job, we would not be very enlightened! We will construct shorter intervals, with the consequence that the true population percentage won't always be in the interval.

II. SAMPLING DISTRIBUTIONS

Two different samples from a population most likely will not have exactly the same sample proportion. The activities in this section teach you about the *sampling distribution*, which describes the variability among repeated samples from the same population. You will learn how to approximate a sampling distribution through simulation. All the work in this section deals with populations for which we know the true percentage of *yeses*.

Application 2

Tossing Four Coins

We know that about 50% of the student population are girls. Suppose that by random sampling we obtain a sample of 4 students and observe whether there are 0, 1, 2, 3, or 4 girls. Toss four coins to simulate the results we are likely to get from this survey. Let heads correspond to *girl* and tails to *boy*. (If you prefer, you can use a different random device, such as rolling four dice, with 1, 2, or 3 corresponding to *girl*.)

1. Toss four coins all at once (or one coin four times).

    a. How many heads did you get?

    b. What is the sample proportion of *girls* (heads)?

    c. Will you get this same sample proportion each time you toss four coins?

2. Now toss the four coins all at once a total of 40 times (giving 40 *trials*). Tally your results on a table like this one. (Several students may want to form a group and combine results to produce a total of 40 trials.)

Number of Heads	Sample Proportion	Tally	Frequency	Proportion of All Trials
0	$\frac{0}{4} = 0.00$	III	3	$\frac{3}{40} = 0.075$
1	$\frac{1}{4} = 0.25$	HHT III	8	$\frac{8}{40} = 0.20$
2	$\frac{2}{4} = 0.50$	HHT HHT HHT I	16	$\frac{16}{40} = 0.40$
3	$\frac{3}{4} = 0.75$	HHT HHT	10	$\frac{10}{40} = 0.25$
4	$\frac{4}{4} = 1.00$	III	3	$\frac{3}{40} = 0.075$
TOTAL			40	$\frac{40}{40} = 1.00$

3. Combine the frequencies from every group in the class and complete a table like the one for question 2.

4. What is the most likely number of heads?

5. What percentage of the time did your class get 1, 2, or 3 *girls* (heads)?

6. List the 16 ways that four coins can land when tossed. We have listed 3 ways to get you started.

1st Coin	2nd Coin	3rd Coin	4th Coin
H	H	H	H
H	H	H	T
H	H	T	H

7. Look at the chart you completed in question 6. In how many ways can we throw

   a. 0 heads?

   b. 1 head?

   c. 2 heads?

   d. 3 heads?

   e. 4 heads?

8. Use your answers to question 7 to calculate the probability of getting

   a. 0 heads.

   b. 1 head.

   c. 2 heads.

   d. 3 heads.

   e. 4 heads.

9. Complete this sentence using answers from question 8: If we observe four randomly chosen students, the probability of this group containing 1, 2, or 3 girls is ____.

10. Compare your answers to questions 5 and 9. Did the simulation give a reasonably accurate answer?

Tossing Eight Coins

1. In this application we will toss a sample of eight coins. Make a table like the one below for tallying the results. Fill in the sample proportion column.

Number of Heads	Sample Proportion	Tally	Frequency	Proportion of All Trials
0 1 2 3 4 5 6 7 8				
TOTAL		10	10	$\dfrac{10}{10} = 1.00$

2. Each student (or group of students) should toss eight coins (or one coin 8 times) a total of 10 times. Fill in the last three columns of your table.

3. Combine the frequencies from every group in the class and fill in a table like the one for question 1.

Answer questions 4 to 7 using the simulation from question 3.

4. What is the most likely number of heads?

5. Estimate the probability of getting 2, 3, 4, 5, or 6 heads.

6. Complete this sentence: If we observe 8 randomly chosen students, about _____ of the time we will have 2, 3, 4, 5, or 6 girls.

7. Compare the table from question 3 of Application 2 with the one from question 3 here.

   a. Are you more likely to get a sample proportion of exactly 0.50 heads if you toss four coins or if you toss eight coins?

   b. Are you more likely to get a sample proportion of heads between 0.25 and 0.75 if you toss four coins or if you toss eight coins?

8. Are you more likely to get exactly 10 heads from tossing 20 coins, or exactly 50 heads from tossing 100 coins?

9. Are you more likely to get a sample proportion from 0.25 to 0.75 from tossing 20 coins (between 5 and 15 heads), or a sample proportion from 0.25 to 0.75 from tossing 100 coins (between 25 and 75 heads)?

Eight coins can land in 2^8 or 256 ways when tossed. How would you like to spend the next few hours listing all 256 ways in order to calculate exact probabilities? Let's rely on simulation from now on to estimate probabilities!

Using Random Number Tables to Make a Sampling Distribution

About 40% of the American public believe schoolchildren have "too many rights and privileges" (Laramie Sunday *Boomerang*, August 11, 1985). Suppose we plan to choose a sample of 20 Americans. Can you guess how many will say they agree with this statement?

To simulate this example, you could use a physical random device, such as a spinner, that would give a probability of 0.40. Instead, we will use the random number table on pages 90 and 91.

A random number table displays digits 0 through 9 in random order. For a population with 40% *yes*es, we assign four of the digits to *yes* and the other six to *no*. For example, digits 0, 1, 2, and 3 could be *yes*, and digits 4, 5, 6, 7, 8, and 9 could be *no*. (Alternatively, digits 6, 7, 8, and 9 could be *yes*, and digits 0, 1, 2, 3, 4, and 5 could be *no*.) The important thing is to decide, before looking at the random number table, which digits correspond to *yes*. Then we pick an arbitrary point on the random number table to start. One way to pick the starting point is to close your eyes and haphazardly put your finger down on the page.

To obtain a sample of size 20, we look at a sequence of 20 digits, going either left or right or up or down from the starting point. It doesn't matter which direction we go to get the 20 digits, as long as we decide on the direction before looking at the table. We determine how many *yes*es are in our sample by counting the number of digits that are 0, 1, 2, or 3. For example, suppose that our first sample of 20 digits is 84310 76343 64238 59419. Then there are 8 *yes*es in the sample. To draw a second sample of size 20, we continue using the next 20 digits. When we get to the edge of the page of random digits, we continue backward in an adjacent row or column.

1. Construct a table like the one on page 9. You will use this table to tally the results for random samples of size 20. Fill in the sample proportion column.

2. Use the random number table to draw a sample of size 20 from a population with 40% *yes*es. Enter a tally mark in your table.

3. Draw 9 more samples and tally them, giving you a total of 10 trials. Then fill in the two right columns in the table.

4. Combine your results with those of other class members in a similar table. (Now the number of trials will be 10 times the number of students.)

Use the table from question 4 to answer questions 5 through 8.

5. What was the smallest number of *yes*es in any one sample?

6. What was the largest number of *yes*es in any one sample?

7. What is the most likely sample proportion of *yes*es?

Number of Yeses	Sample Proportion	Tally	Frequency	Proportion of All Trials
0				
1				
2				
3				
4				
5				
6				
7				
8				
9				
10				
11				
12				
13				
14				
15				
16				
17				
18				
19				
20				
TOTAL		**10**	**10**	**1.00**

8. Suppose that the newspaper report is correct: 40% of the American public believe schoolchildren have too many rights and privileges. If you have a random sample of 20 Americans, then make the following estimates.

   a. Estimate the probability that exactly 8 people will believe schoolchildren have too many rights and privileges.

   b. Estimate the probability that 6 or fewer people will believe it.

   c. Estimate the probability that the sample proportion believing the statement will be from 0.30 to 0.50, inclusive (that is, the probability that from 6 to 10 people will believe it).

   d. Estimate the probability that all 20 Americans will believe the statement.

You just approximated a *sampling distribution* through simulation—specifically, the sampling distribution of the number of *yes*es in a sample of size 20 drawn from a population with 40% *yes*es. In Applications 2 and 3, you constructed sampling distributions for the number of heads in samples of size 4 and 8 from a population with 50% heads.

The sampling distribution shows the amount of variability from one random sample to another from a specific population. To construct a sampling distribution, we must know both the population percentage and the sample size. A sampling distribution can be used in either of two equivalent forms: the number of *yes*es in a sample, or the sample proportion. For example, using the sampling distribution for a sample of size 20 from a population with 40% *yes*es, we can use the "proportion of all trials" column to estimate the probability of 3 *yes*es, or equivalently the probability of a sample proportion of 0.15.

It is often possible to construct an exact sampling distribution using probability formulas. Using simulation you may obtain a sampling distribution slightly off from the exact one. However, the more trials you run in the simulation, the closer your approximated sampling distribution should be to the exact one. In this book, we will not discuss the probability formulas for deriving the exact sampling distribution; we will always use simulation to approximate the sampling distribution.

9. Describe how the sampling distribution from 10 trials (question 3) differs from the sampling distribution from many more trials (question 4). Which sampling distribution do you think is closer to the one calculated from probability formulas?

III. BOX PLOTS FROM SAMPLING DISTRIBUTIONS

You have used simulation to construct sampling distributions, and you have used tables like the one below to describe these distributions. This table was constructed using samples of size 20 from a population containing 40% *yes*es. We did 40 trials. Next you will learn how to use a 90% box plot to summarize this sampling distribution.

Number of Yeses	Sample Proportion	Frequency	Proportion of All Trials
0	0.00	0	0
1	0.05	0	0
2	0.10	0	0
3	0.15	0	0
4	0.20	1	0.025
5	0.25	3	0.075
6	0.30	4	0.10
7	0.35	8	0.20
8	0.40	9	0.225
9	0.45	1	0.025
10	0.50	5	0.125
11	0.55	8	0.20
12	0.60	1	0.025
13	0.65	0	0
14	0.70	0	0
15	0.75	0	0
16	0.80	0	0
17	0.85	0	0
18	0.90	0	0
19	0.95	0	0
20	1.00	0	0
TOTAL		40	1.00

On page 2, you read a statement by the Gallup poll that "one can say with 95% confidence . . . ." Gallup uses 95% box plots. We will use 90% instead of 95% because the computations necessary to make a box plot are easier with 90%.

Displaying sampling distributions in a plot makes it easier to analyze and compare them. Thus, we will use box plots to focus attention on the most important features of the sampling distributions.

The following figure is a 90% box plot of the sample proportions in this example. The number line at the top is for sample proportions and goes from 0.0 to 1.0, as in the second column of the preceding table. The number line at the bottom of the plot is the corresponding number of *yeses* in the sample, from the first column of the table. We have positioned the box along the number line to represent the frequencies of these sample proportions in the 40 trials (the third and fourth columns of the table). Next we will learn how to construct this 90% box plot.

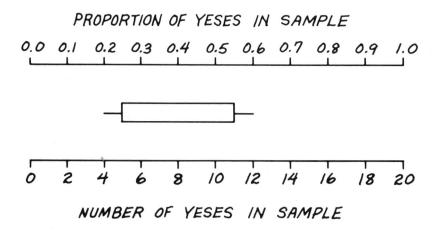

We want to have the sample proportions from the middle 90% of the 40 trials lie inside the box, including the edges. We also want to have the sample proportions from 5% of the 40 trials lie in the lines on either side of the box, the *whiskers*. Ninety percent of 40 is 36. Thus, we want 36 of the observed sample proportions in the box and the remaining 4 in the whiskers (2 in each). Look at the frequency column in our table. Because the whiskers must contain the two smallest sample proportions and the two largest ones, we draw our box starting with the *third* smallest sample proportion (0.25) and extend it to the *third* largest sample proportion (0.55). We then draw the whiskers to represent the remaining 10% of sample proportions: One extends to the left to the smallest recorded sample proportion, 0.20, and the other extends to the right to the largest recorded sample proportion, 0.60.

Because of ties in these sample proportions, we cannot get exactly 36 values in the box. This box actually contains 38 values (including the edges). When ties occur, we will always construct the box so that no more than 5% of the values are in either whisker. One whisker might contain fewer than 5% but never more. Similarly, with ties we might have to put more than 90% of the sample proportions in the box, but we will never put in fewer than 90%.

Application 5

Constructing a 90% Box Plot

We took 100 random samples, each of size 20, from a population with 50% *yes*es and got these results:

Number of Yeses	Sample Proportion	Frequency
0	0.00	0
1	0.05	0
2	0.10	0
3	0.15	0
4	0.20	1
5	0.25	2
6	0.30	5
7	0.35	12
8	0.40	11
9	0.45	10
10	0.50	16
11	0.55	21
12	0.60	8
13	0.65	8
14	0.70	4
15	0.75	1
16	0.80	1
17	0.85	0
18	0.90	0
19	0.95	0
20	1.00	0
TOTAL		**100**

1. a. When you make a 90% box plot from 100 trials, how many sample proportions should the box ideally contain?

    b. How many sample proportions should each whisker ideally contain?

    c. What is the sixth smallest sample proportion from the 100 trials?

    d. What is the sixth largest sample proportion from the 100 trials?

    e. Make the 90% box plot of the sample proportions from the 100 trials.

    f. How many sample proportions actually ended up in the box?

2. We took 200 random samples, each of size 10, from a population with 50% *yes*es and got these results:

Number of Yeses	Sample Proportion	Frequency
0	0.00	0
1	0.10	5
2	0.20	10
3	0.30	21
4	0.40	42
5	0.50	47
6	0.60	39
7	0.70	26
8	0.80	9
9	0.90	1
10	1.00	0
TOTAL		200

Make a 90% box plot of these sample proportions. The box will start at the 11th sample proportion from each end.

3. Review your work for Application 4 (page 8). For question 4, you constructed a sampling distribution for samples of size 20 from a population with 40% *yes*es and with the number of trials equal to 10 times the number of students in your class.

   a. To construct a 90% box plot, how many sample proportions must you count in from either end to determine the edges of the box?

   b. Construct the 90% box plot for your sampling distribution.

Application 6

Deciding If the Sample Proportion is Likely or Unlikely

We have constructed the 90% box plot so that it contains the sample proportions from the middle 90% of the trials. We will call the sample proportions inside the box (including its edges) the *likely sample proportions*, because most of the trials (specifically, 90% of them) gave one of these sample proportions. The lines on either side of the box, the whiskers, represent sample proportions from the remaining 10% of the trials, with 5% in each whisker. We call the sample proportions falling in the whiskers the *unlikely sample proportions*.

If you take further samples, you might even get a sample proportion that is outside the whiskers. We also call such sample proportions unlikely. Thus, unlikely sample proportions can fall either in the whiskers or outside the whiskers. You are very unlikely to get a sample proportion outside the whiskers, however.

For example, using the 90% box plot of the sampling distribution for a sample of size 20 from a population with 40% *yes*es, we see that a sample proportion of 0.50 (10 *yes*es out of 20) is a likely sample proportion. However, a sample proportion of 0.60 (12 *yes*es out of 20) is an unlikely sample proportion. Use this 90% box plot to answer the following questions.

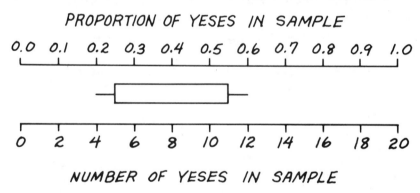

PROPORTION OF YESES IN SAMPLE

NUMBER OF YESES IN SAMPLE

1. Forty percent of Irish voters voted to lift a constitutional ban on divorce in cases of long-term separation (Newark *Star-Ledger*, June 28, 1986). If you take a random sample of 20 Irish voters, is getting 9 (sample proportion of 0.45) who voted this way a likely or unlikely sample proportion?

2. If you take a random sample of size 20 from a population with 40% *yes*es, will each sample proportion below be a likely or unlikely sample proportion?

   a. 0.40

   b. 0.65

   c. 0.20

   d. 0.90

   e. 0.35

3. For a random sample of size 20 from a population with 40% *yeses*, tell whether each result below gives a likely or unlikely sample proportion.

   a. 20 *yes*, 0 *no*   d. 8 *yes*, 12 *no*

   b. 12 *yes*, 8 *no*   e. 4 *yes*, 16 *no*

   c. 10 *yes*, 10 *no*   f. 0 *yes*, 20 *no*

4. Forty percent of all plain M&M's are brown. If you take a random sample of 20 M&M's, tell whether each number of brown M&M's below is likely or unlikely.

   a. 9 brown

   b. 2 brown

   c. 15 brown

   d. 7 brown

5. The U.S. Bureau of Labor Statistics reports that about 40% of all women with children under the age of 18 do not work. Suppose that you select a random sample of 20 women with children under the age of 18 and ask each woman whether she works. List the likely sample proportions.

6. Complete this sentence:
   If we take a random sample of size 20 from a population with 40% *yeses*, 90% of the time we will get a sample proportion of *yeses* between _____ and _____ .

7. According to *On Campus*, the official publication of the American Federation of Teachers, a 1983 Gallup survey found that 40% of the American public favors a longer school year (10 months). Suppose that you select a random sample of 20 Americans and learn that 4 favor a longer school year. If Gallup is right, is 4 out of 20 a likely or unlikely sample proportion? Given this sample proportion, would you think Gallup is right?

8. According to the 1980 U.S. census, about 40% of the population of the city of Chicago is black (*World Almanac*, 1984). In a random sample of 20 Chicagoans, will each result below give a likely or unlikely sample proportion?

   a. all are black

   b. half are black

   c. 12 are black

   d. 30% are black

9. (*Optional*) Ask 20 adults this question: "Do you favor a longer school year?" Do you think your sample is representative of the American public? Why or why not? Is your sample proportion likely or unlikely if Gallup is right (question 7)? On the basis of your survey, do you think Gallup is wrong? Why or why not?

Making and Interpreting the 90% Box Plot for a Population with 80% Yeses

About 80% of U.S. adults favor graduation exams even if failure to pass the test could deprive their children of a regular high school diploma (*USA Today*, April 1, 1985).

1. Construct a table like the following one. Fill in the sample proportion column. Then, using a random number table, your class should draw samples of size 20 from a population with 80% *yeses*. Continue until you have 40 trials, and enter the results in the two right columns of the table.

Number of Yeses	Sample Proportion	Tally	Frequency
0			
1			
2			
3			
4			
5			
6			
7			
8			
9			
10			
11			
12			
13			
14			
15			
16			
17			
18			
19			
20			
TOTAL		40	40

2. Make a 90% box plot of the sample proportions.

3. What percentage of your trials actually ended up inside your box, including the edges? (Your answer must be 90% or larger.)

Use the 90% box plot from question 2 to answer questions 4 through 10.

4. If we ask a random sample of 20 U.S. adults if they favor graduation exams, are the following results likely or unlikely?

   a. 20 *yes*, 0 *no*   d. 10 *yes*, 10 *no*

   b. 18 *yes*, 2 *no*   e. 5 *yes*, 15 *no*

   c. 14 *yes*, 6 *no*

5. About 80% of U.S. adults support the right of school authorities to open school lockers or examine personal property for drugs, liquor, or other contraband (Laramie Sunday *Boomerang*, August 11, 1985). Is it likely or unlikely that a poll of 20 randomly selected adults would show

   a. just 2 favoring this practice?

   b. all 20 favoring it?

   c. 17 favoring it?

   d. 16 favoring it?

6. According to a 1979 census of inmates of juvenile detention and correctional facilities, 80% of those under correctional supervision were male (U.S. Department of Justice, *Report to the Nation on Crime and Justice*, 1983). If we take a random sample of 20 such inmates, is it likely or unlikely that 15 will be male?

7. Complete this sentence:
   If we draw a random sample of size 20 from a population with 80% *yes*es, we estimate that the proportion of *yes*es in our sample will be from _____ to _____ at least 90% of the time.

8. In a random sample of 20 adults, 10 favor graduation exams. Is this a likely sample proportion if 80% of all adults favor graduation exams?

9. About 80% of Americans are against paying higher taxes for defense (*New York Times*, April 4, 1984). If we obtain a random sample of 20 Americans and ask each person if he or she is against paying higher taxes for defense, what are the likely sample proportions?

10. A teacher thought that 80% of the students in his school had seen *E.T.*, but when he asked 20 students at random, he learned that 14 had seen this movie. Do you think he was wrong? Why or why not?

IV. CHARTS OF 90% BOX PLOTS

You have made 90% box plots of sampling distributions for random samples of size 20, including one for populations with 40% *yes*es and one for populations with 80% *yes*es. Here are these box plots placed next to each other:

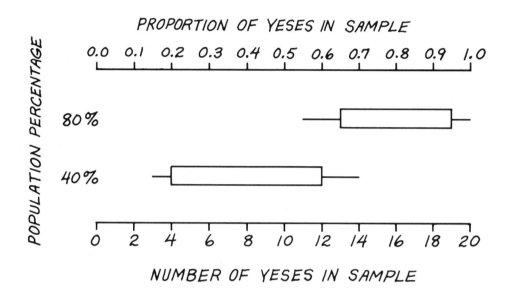

To make these 90% box plots as accurate as possible, we constructed them (using a computer) with many more trials than you did. Thus, your 90% box plots might differ a bit from ours. (They shouldn't differ too much, however.)

We have also made 90% box plots for random samples of size 20 from populations with 5%, 10%, 15%, on up to 100% *yes*es. These box plots, all placed in order next to each other, are on page 92. We constructed these 90% box plots using many trials as well. Note that for a population with 0% *yes*es every sample will, of course, have 0 *yes*es; the sample proportion is always 0.0. Thus, we drew this box as a simple vertical bar at the sample proportion of 0.0. Similarly, for a population with 100% *yes*es, every sample will contain all *yes*es, so we drew the box for the sampling distribution as a vertical bar at sample proportion 1.0.

Remember, each row on the chart of 90% box plots represents a different population, with the indicated percentage of *yes*es. You will find two kinds of questions in the following applications. Given information about the population, you must first find that population in the left column of the chart, then use the information from its box plot to answer the question. Alternatively, if you have information about the sample, you must first find that sample proportion along the scale at the top, then read down to find out the population or populations that answer the question.

Application 8

Reading Charts of 90% Box Plots

Use the chart of 90% box plots on page 92 to answer the questions below. For questions 1 through 6, first find the appropriate population percentage to the left or right of the chart. Then see if the given sample proportion is in the box for that population percentage. If so, this sample proportion is likely. If this sample proportion is in the whisker or outside the box plot, it is unlikely.

1. A sample of size 20 is selected randomly from a population with 45% *yes*es. Is each of the following results likely or unlikely? Remember that sample proportions falling on the edge of the box are considered inside it and so are likely.

   a. a sample proportion of 0.10 *yes*es

   b. a sample proportion of 0.20 *yes*es

   c. a sample proportion of 0.50 *yes*es

   d. a sample proportion of 0.75 *yes*es

2. A sample of size 20 is selected randomly from a population with 10% *yes*es. Is each of the following results likely or unlikely?

   a. a sample proportion of 0.00 *yes*es

   b. a sample proportion of 0.20 *yes*es

   c. a sample proportion of 0.50 *yes*es

3. About 45% of all mathematicians in the United States are women (*Los Angeles Times*, March 7, 1984). If we take a random sample of 20 mathematicians, are the following results likely or unlikely?

   a. 10 women and 10 men

   b. 15 women and 5 men

   c. 8 women and 12 men

4. Imagine you are taking a true–false test about the Byzantine civil service system. For each of the 20 questions, you discreetly flip a coin and answer *true* if the coin lands heads and *false* if it lands tails. Are each of the following results likely or unlikely?

   a. a 100% score on the test

   b. a 90% score on the test

   c. a 80% score on the test

   d. a 70% score on the test

   e. What scores are you likely to get?

5. According to the National Center for Education Statistics, 30% of male high school seniors have taken trigonometry, compared with 22% of the female students (*Los Angeles Times*, March 7, 1984). If you take a random sample of 20 male high school seniors and ask each if he has taken trig, what are the likely sample proportions?

6. Sixty-five percent of men are fully or partially bald by the time they reach age 55 (*Los Angeles Times*, December 9, 1983). If you check 20 randomly selected 55-year-old men for baldness, what are the likely sample proportions?

   To answer the remaining questions, you must first find the appropriate sample proportion across the top of the chart, or the corresponding number of *yes*es at the bottom. Then read down or up to see the population percentages for which this sample proportion is likely.

7. A random sample of size 20 contains a sample proportion of 0.20 *yes*es. For which of the following population percentages is this a likely sample proportion?

   a. one with 5% *yes*es   f. one with 30% *yes*es

   b. one with 10% *yes*es   g. one with 35% *yes*es

   c. one with 15% *yes*es   h. one with 40% *yes*es

   d. one with 20% *yes*es   i. one with 45% *yes*es

   e. one with 25% *yes*es   j. one with 50% *yes*es

8. A random sample of size 20 contains a sample proportion of 0.50 *yes*es. For which population percentages is this a likely sample proportion?

9. A random sample of size 20 contains 14 *yes*es. For which population percentages is this a likely sample proportion?

10. A random sample of size 20 contains 20 *yes*es. For which population percentages is this a likely sample proportion?

Reading Charts of 90% Box Plots for Samples of Size 100

In this application, the samples will be of size 100. Consequently, you can no longer use the chart on page 92, which was constructed from samples of size 20. Our computer has made a similar chart from samples of size 100. You will find it on page 95. Use it to answer the following questions.

For questions 1 through 5, first find the appropriate population percentage to the left or right of the chart. Then see if the given sample proportion is in the box for that population percentage.

1. For a random sample of size 100 from a population with 55% *yeses*, state whether the following are likely or unlikely results.

   a. a sample proportion of 0.90 *yeses*

   b. a sample proportion of 0.70 *yeses*

   c. a sample proportion of 0.50 *yeses*

   d. a sample proportion of 0.20 *yeses*

2. Ninety percent of U.S. adults agree with the recommendation that high school students take three years of math (*USA Today*, April 1, 1985). Assuming that this suggestion is no April Fool's joke, is it likely or unlikely that a poll of 100 randomly selected adults would show

   a. 100 *yeses* and 0 *nos*?

   b. 92 *yeses* and 8 *nos*?

   c. 84 *yeses* and 16 *nos*?

   d. 62 *yeses* and 38 *nos*?

   e. 40 *yeses* and 60 *nos*?

3. About 25% of Americans bite their fingernails (*Los Angeles Times*, December 9, 1983). If you select a random sample of 100 Americans and check each one for nail biting, what are the likely sample proportions?

4. About 5% of Americans find life dull (*Los Angeles Times*, April 13, 1986). If you ask a random sample of 100 Americans if they find life dull, what are the likely sample proportions?

5. The Census Bureau reports that about 15% of the adults living in the U.S. are illiterate in English (*Cincinnati Enquirer*, April 21, 1986). What are the likely sample proportions if we check a random sample of 100 adults living in the U.S. for illiteracy in English?

To answer questions 6 through 11, first find the appropriate sample proportion across the top of the chart, or the corresponding number of *yeses* at the bottom. Then read down or up to see for which population percentages this sample proportion is likely.

6. A random sample of size 100 contains a sample proportion of 0.20 *yeses.* For which of the following population percentages is this a likely sample proportion?

   a. one with 5% *yeses*

   b. one with 10% *yeses*

   c. one with 15% *yeses*

   d. one with 20% *yeses*

   e. one with 25% *yeses*

   f. one with 30% *yeses*

   g. one with 35% *yeses*

   h. one with 40% *yeses*

   i. one with 45% *yeses*

   j. one with 50% *yeses*

7. A random sample of size 100 contains 50 *yeses.* For which population percentages is this a likely sample proportion?

8. A random sample of size 100 contains 89 *yeses.* For which population percentages is this a likely sample proportion?

9. A study of about 100 divorced couples with children found that 24 were "fiery foes" who rarely communicated (*USA Today*, April 23, 1986). For which population percentages is this a likely sample proportion?

10. A sample (selected at random, we hope) of 100 lower- and middle-class boys found that 8 had conduct disorders, such as stealing, fighting, and running away from home (*Los Angeles Times*, November 25, 1982). For which population percentages is this a likely sample proportion?

11. The same study also investigated about 100 lower- and middle-class boys with symptoms of hyperactivity. Of the boys in this sample, a proportion of 0.27 had conduct disorders. For which population percentages is this a likely sample proportion?

12. Considering your answers to questions 10 and 11, do you think hyperactive boys are more likely than typical boys to have conduct disorders? Explain.

13. Compare the 90% box plots for random samples of size 100 with those for samples of size 20.

    a. Which chart has shorter box plots?

    b. Why do you think this chart has shorter box plots?

Reviewing Charts of 90% Box Plots

Use the box plots on page 93, which were constructed from samples of size 40, to answer the following questions.

1. Suppose you reach in a jar of marbles, pull out 40, and find that 18 are blue.

   a. Is this sample proportion likely if 60% of the marbles in the jar are blue?

   b. For which population percentages is a sample proportion of 18 out of 40 likely?

2. Ninety-five percent of U.S. adults believe that students should pass math and reading tests before they graduate from high school (*USA Today*, April 1, 1985). If we take a random sample of 40 U.S. adults,

   a. is it likely or unlikely that a sample proportion of 0.875 will approve of this requirement?

   b. is it likely or unlikely that all 40 will approve?

3. What are the likely sample proportions if we draw a random sample of size 40 from a population with 25% *yes*es?

4. For which population percentages is a sample proportion of 0.25 from a random sample of size 40 a likely result?

5. Alcohol was found in 70% of the blood samples taken from male drivers, age 15 to 34, who died in motor vehicle crashes in four California counties in 1982–83 (*Public Health Reports*, 1985). If we were to take a random sample of 40 such drivers, would we be likely to find

   a. a sample proportion of 0.575 with alcohol in their blood?

   b. 33 drivers with alcohol in their blood?

6. According to General Mills, about 90% of Americans eat breakfast at least some of the time (*Los Angeles Times*, December 9, 1983). If we select a random sample of 40 Americans and ask each person whether he or she eats breakfast at least some of the time, what are the likely sample proportions?

V. CONFIDENCE INTERVALS

This section contains the central idea of this book. We will put together everything we have learned so far and will be able to make statements like this one:

> I took a random sample of 20 students at my school and asked them if they love math. Because 6 of them said *yes*, I am fairly sure that if I ask all students at my school this question, between 15% and 50% will say *yes*. However, for every 100 times that I give such an interval, I expect to be right 90 times and wrong 10 times.

Let's see how to construct the interval we've just described.

Suppose we get 6 *yes*es in a random sample of size 20, for a sample proportion of 0.30. From the chart of 90% box plots on page 92 for samples of size 20, we see that this result is likely from populations with 15%, 20%, 25%, 30%, 35%, 40%, 45%, and 50% *yes*es. We say that 15% to 50% is a **90% confidence interval**. We think that the population has between 15% and 50% *yes*es.

To find the 90% confidence interval for the percentage of *yes*es in a population, lay a ruler down the column giving the sample proportion, as in the diagram below. The line will intersect some but not all of the boxes. If the line falls exactly on the edge of the box, we say that the line intersects this box. The boxes intersected by the line represent the populations for which the sample proportion is likely. Thus, these populations make up the 90% confidence interval.

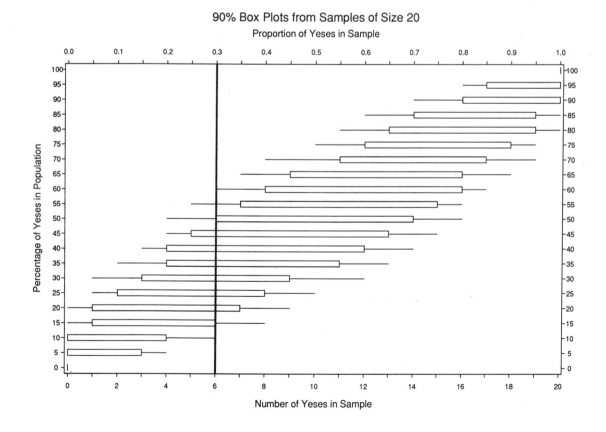

90% Box Plots from Samples of Size 20
Proportion of Yeses in Sample

Let's do another example. Suppose we get 14 *yes*es in a random sample of size 20, for a sample proportion of 0.70. Laying a ruler down the 0.70 column, we find that the ruler intersects the boxes from the 50% to the 85% populations. Our sample is a likely result from populations with 50% to 85% *yes*es. So our 90% confidence interval for the percentage of *yes*es in the population is 50% to 85%.

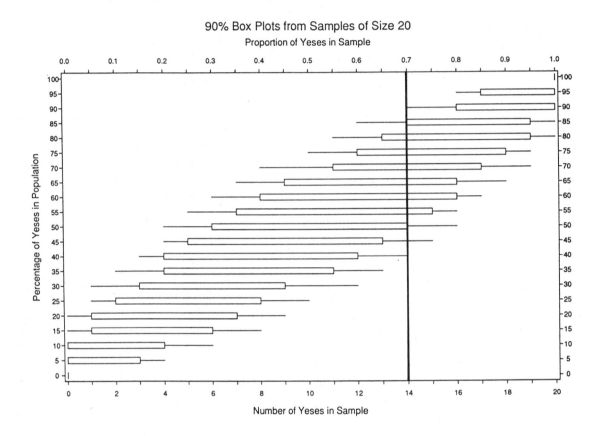

90% Box Plots from Samples of Size 20

Further, when we make statements like "The percentage of *yes*es in the population is between 50% and 85%," we will be right 90% of the time and wrong 10% of the time. This last statement is not as obvious as it might seem. You will learn more about this statement through the simulations in Applications 12 and 13, and Application 14 gives the mathematical argument underlying this statement.

Application 11

Finding Confidence Intervals

1. Your teacher has the container of objects you used at the beginning of this book. Draw a random sample of size 20 and find the 90% confidence interval for the percentage of *yeses* in the container. Does your confidence interval contain the true percentage of *yeses*?

2. *Penny Power* magazine gave each of 20 eighth graders three erasable pens and a nonerasable Bic Stic ballpoint pen. Each pen had a medium point and blue ink. The students used the pens for one week while doing their usual schoolwork and homework. At the end of the week, 14 students preferred the Bic Stic to any of the erasables (*Penny Power*, August/September 1984). Assuming that the magazine selected the students and pens randomly, find the 90% confidence interval for the percentage of all eighth graders who prefer the Bic Stic to these erasables.

3. Of the 20 students in question 2, 11 chose the Scripto Erasable as the best of the erasable pens.

   a. What is the 90% confidence interval for the percentage of all eighth graders who prefer the Scripto Erasable to the other erasables?

   b. Can you be fairly confident that at least half of all eighth graders prefer this erasable pen? Why or why not?

4. In a study of advanced chronic multiple sclerosis (MS), 20 patients spent 30 hours in a high-pressure oxygen chamber with 10% oxygen and 90% nitrogen. Only one patient improved (*Los Angeles Times*, January 27, 1983). Find the 90% confidence interval for the percentage of MS patients who will improve with this treatment.

5. In another study of MS patients, 16 of 20 stabilized or improved after treatment with anticancer and steroid drugs to suppress their immune systems (*Los Angeles Times*, January 27, 1983). Find the 90% confidence interval for the percentage of MS patients who will improve or stabilize with this treatment.

6. To improve health care for premature babies, physicians wanted to learn which of three types of milk would give the best results. A Duke University pediatrician studied 60 premature babies who weighed 3 pounds or less. In this sample, 20 babies were fed milk from mothers who had had premature babies, 20 were fed milk from mothers who had had full-term babies, and the remaining 20 were fed formula. By the sixth week of feeding, 18 of the babies on the formula, 17 on the preterm milk, and 12 on full-term milk had gained normal weights. However, several babies in the study unexpectedly became sick. Six babies on formula became sick and 2 of them died. One baby on the full-term milk died. None on the preterm milk died (*Los Angeles Times*, February 3, 1983).

   a. Suppose you were a pediatrician associated with this study. What is the single most important feature of the data you would investigate first?

   b. What is the 90% confidence interval for the percentage of babies on full-term milk who will regain normal weight?

   c. What is the 90% confidence interval for the percentage of babies on formula who will regain normal weight?

   d. What is the 90% confidence interval for the percentage of babies on preterm milk who will regain normal weight?

   e. What overall conclusions would you make if you were a pediatrician?

Estimating the Percentage of Digits That Are Even

1. Use the random number table on pages 90 and 91 to get a random sample of 20 digits. (Each student should obtain a different random sample.)

2. What is the number of even digits in your sample? Remember that 0 is an even digit!

3. Using the proportion of even digits in your sample and the chart of 90% box plots on page 92, find the 90% confidence interval for the percentage of even digits in a random number table.

4. What is the true percentage of even digits in a huge list of random numbers?

5. Does your 90% confidence interval contain the true percentage?

6. What percentage of the students in your class do you think will answer *yes* to question 5?

7. Determine the percentage of students in your class who did answer *yes* in question 5. Is this percentage about what you expected?

8. Complete this sentence:
   If 100 students do the experiment described in questions 1 through 5, about _____ of them will answer *yes* to question 5.

9. Complete this sentence:
   If 200 students do the experiment described in questions 1 through 5, about _____ of them will answer *yes* to question 5.

Determining How Often the Population Percentage Is in the Confidence Interval

The data sheet on page 89 shows 12 arrays of X's and O's. Each array contains 10 samples of size 20 drawn from some population. An X is a *yes* and an O is a *no*. Select two or three students to work with each array.

1. The first row of your array is your first sample of size 20.

    a. Count the number of X's in this row.

    b. Find the sample proportion of *yes*es in this row.

    c. Use the sample proportion of *yes*es and the chart of 90% box plots on page 92 to find the 90% confidence interval for the percentage of *yes*es in your population.

2. The second row of your array is your second sample of size 20. Repeat question 1 for the second row and then for each of the eight remaining rows. Complete all but the last column of a chart like the one below.

Row	Number of X's	Sample Proportion of X's	90% Confidence Interval for the Population Percentage of X's	Is the True Population Percentage in the Confidence Interval?
1				
2				
3				
4				
5				
6				
7				
8				
9				
10				

3. You now have 10 confidence intervals. How many of them do you expect to contain the true population percentage?

4. Your class constructed 120 confidence intervals altogether. How many of these confidence intervals do you expect to contain the true population percentage?

5. Ask your teacher for the true population percentages.

    a. Fill in the last column of your chart. How many of your 10 confidence intervals contain the true percentage for your population?

    b. What percentage of the 120 intervals constructed in your class contain their true population percentages?

Now let's see how to answer the question suggested by the title of this application. How often will our confidence interval contain the true population percentage? Your answer to question 5 gives an estimate based on the simulations in your class. Similarly, your answer to question 7 of Application 12 also gives an answer based on different simulations done by your class. Both answers should be about the same. You might expect them to be about 90%, because we are using 90% box plots. Most classes will find, however, that the confidence intervals they have constructed contain the true population percentage a little more than 90% of the time because the boxes actually contain slightly more than 90% of the possible samples (see page 12).

In these two applications, you have been able to check whether each confidence interval included the true population percentage because we know what the true population percentages are. However, in a real survey, we do not know the true population percentage. (If we did know, we'd have no reason to take a random sample to get an estimate!) For real surveys, the true population percentage will either be in the confidence interval or it will not; we never know which. All we can say is that we expect that 90% of the confidence intervals constructed using our method will contain the true population percentage.

Thus, simulations give one way to answer our question. Using mathematical reasoning is a different way to learn that about 90% of our confidence intervals will contain the true population percentage. Application 14 explains this mathematical argument, which does not use simulation.

Analyzing Why 90% of Confidence Intervals
Contain the Population Percentage

In this activity, you will see why you can expect 90% of all confidence intervals you construct to contain the true population percentage. As we said on the first page of this section, this conclusion is not obvious. We know that the sample proportion will be in the box 90% of the time. Why does this fact imply that 90% of all confidence intervals will contain the population percentage?

The following discussion and questions 1 through 7 assume that we obtain the sample proportion for a random sample from a population with 20% *yeses*. The figure below is a simplified chart of 90% box plots that we will use to analyze confidence intervals. The population labeled ✳ has an unknown percentages of *yeses*; we will use it in question 9 of this application.

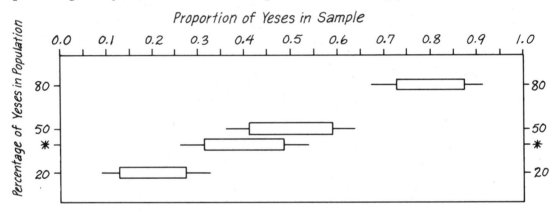

Let's use the above picture to consider what the confidence interval would look like. First, suppose that the sample proportion falls to the *right* of the box for the population with 20% *yeses*, say at X on the picture as shown below.

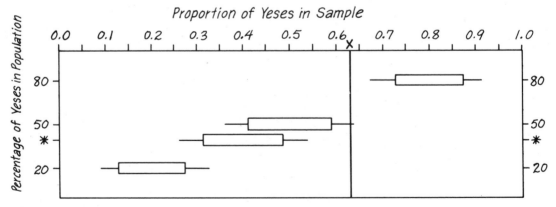

To construct the confidence interval from this sample proportion, we look at a vertical line down from X and see which boxes it intersects. In the above picture, the line clearly does not intersect the box for the population with 20% *yeses*.

1. If the sample proportion X falls to the right of the box for the population with 20% *yes*es, will the resulting confidence interval contain 20%? In other words, will the confidence interval contain the true population percentage?

2. Suppose the sample proportion X falls to the left of the box for the population with 20% *yes*es. Will the confidence interval contain 20%? In other words, will the confidence interval contain the true population percentage?

3. Now suppose that the sample proportion falls inside this box. Draw a schematic figure, similar to the preceding ones, to represent this situation.

4. When the sample proportion falls inside this box, will the confidence interval contain 20%?

5. What is the probability that the sample proportion will fall inside the box for the population with 20% *yes*es?

6. Now put together your answers to questions 4 and 5. With random sampling from a population with 20% *yes*es, what is the probability of getting a sample proportion whose confidence interval contains 20%? In other words, what is the probability of getting a confidence interval that contains 20%?

7. What is the probability of getting a confidence interval that does not contain 20%? In what situation will the confidence interval not contain 20%?

8. Now suppose that we obtain the sample proportion for a random sample from a population with 50% *yes*es. What is the probability of getting a sample proportion whose confidence interval contains 50%? (Note that this question is like question 6, but with 50% *yes*es rather than 20%. To answer question 8, you might find it helpful to think through questions 1 through 7, replacing every 20% with 50%.)

9. Now suppose that we obtain the sample proportion for a random sample from a population with some unknown percentage of *yes*es. Call this unknown percentage *, as indicated in the figure on page 32. What is the probability of getting a sample proportion whose confidence interval contains the percentage *?

This analysis applies to every population, whether its percentage of *yes*es equals 20%, 50%, 80%, or *%. In brief, no matter which population a random sample comes from, the sample proportion will be in the box for that population 90% of the time. The sample proportions in the box in turn produce confidence intervals that include this population percentage. Thus, when the sample proportion is in the box, the confidence interval includes the true population percentage. But we know that the sample proportions will be in the box about 90% of the time, so the confidence intervals will also include the true (but unknown) population percentage about 90% of the time.

Working with Different Sample Sizes

So far we have mainly used sample sizes of 20 and 100. This application will show you the effect of different sample sizes on the length of the confidence interval. To answer the questions below, use the charts of 90% box plots for samples of sizes 20, 40, and 80 on pages 92, 93, and 94.

1. Are you likely or unlikely to get a sample proportion of 0.50 *yes*es from a population with 35% *yes*es if the sample size is

    a. 20?

    b. 40?

    c. 80?

2. Are you likely or unlikely to get a sample proportion of 0.70 *yes*es from a population with 80% *yes*es if the sample size is

    a. 20?

    b. 40?

    c. 80?

3. Which sample size has the longest box plots?

4. Which sample size has the shortest box plots?

5. Do larger sample sizes have longer or shorter box plots? Why?

6. A random sample of size 20 contains 10 *yes*es. Find the 90% confidence interval for the percentage of *yes*es in the population.

7. A random sample of size 40 contains 20 *yes*es. Find the 90% confidence interval for the percentage of *yes*es in the population.

8. A random sample of size 80 contains 40 *yes*es. Find the 90% confidence interval for the percentage of *yes*es in the population.

9. Look at your answers to questions 6, 7, and 8, and complete this sentence:
   As the sample size increases, the length of the confidence interval
   _____ .

10. If we were to compare the lengths of confidence intervals for random samples of size 300, 600, and 1200, which one would be shortest?

11. *True or False*: With larger sample sizes, the sample proportion is more likely to be close to the population percentage.

12. To decrease the length of the confidence interval, must a pollster increase or decrease the sample size? Why might he or she choose not to do this?

Application 16

Reviewing Confidence Intervals

1. Assume we select a random sample of size 20 and obtain 13 *yes*es. Is this sample proportion likely or unlikely if the entire population contains 70% *yes*es?

2. According to the 1980 U.S. census, 30% of males 15 years and over have never been married. If we take a random sample of 20 males 15 years and older and ask if they are single (never married), what are the likely sample proportions?

3. If we get 13 *yes*es in a random sample of size 20, what is the 90% confidence interval for the percentage of *yes*es in the population?

4. Assume we take a random sample of size 20 from a population with 10% *yes*es. Are we likely or unlikely to get

   a. 5 *yes*es and 15 *no*s?

   b. all *no*s?

   c. a sample proportion of 0.30?

5. Assume we select a random sample of size 20 and get a sample proportion of 0.90 *yes*es. Is this sample proportion likely from a population with 75% *yes*es?

6. If we take a random sample of size 20 from a population with 25% *yes*es, is 0.40 a likely sample proportion?

7. Find the 90% confidence interval for the percentage of *yes*es in the population if a random sample of size 20 contains 11 *yes*es.

8. About how many of every 300 90% confidence intervals will contain the true population percentage?

9. To decrease the length of the confidence interval, should you increase or decrease the sample size?

10. (*For those who did Application 14*) Explain why 90% confidence intervals contain the population percentage 90% of the time.

VI. METHODS OF SAMPLING

We have used the term *random sample* often in our discussions so far because our method of constructing confidence intervals is legitimate only for samples selected randomly from the population. This section discusses what random sampling (sometimes called *simple random sampling*) is and why it is important.

Definition of Random Sample

A sample is random if it is selected so that:

1. each member of the population is equally likely to be chosen;

2. the members of the sample are chosen independently of one another.

Note that obtaining a randomly chosen sample depends on *the way in which the sample is drawn*, not on the specific members of the population that happen to end up in the sample.

For example, suppose we want to select a random sample of 20 seniors from a class of 300 at a certain school. We could put the name of each senior on a card, put the cards in a box, mix them up, and draw 20. To see if this method gives random sampling, we must check the two parts of the above definition. For this method,

1. every senior had the same chance of being chosen, and

2. we drew the names independently of each other. (In other words, we didn't let best friends staple their cards together or do anything else that would interfere with drawing cards individually.)

Thus, this selection process is random sampling. A sample selected in this way is a random sample, no matter which specific seniors end up in the sample.

A second way to select a random sample of 20 students from a class of 300 is to use a random number table. We could assign each student a different number from 1 to 300, enter a random number table at some arbitrary location, and take three digits at a time as a random number. The 1000 possible values are 000, 001, 002, . . . , 999, and each is equally likely. If a value from 001 to 300 arises, we put the corresponding student in the sample. If 000 or one of the values from 301 to 999 arises, or if a number repeats, we disregard it and go on to the next random number. We continue until we have 20 students in the sample.

Suppose the 20 names we draw all happen to be members of the girls' softball team. Is this group a random sample? Yes, it is, because we selected it randomly from the population of all seniors. Is this group *representative* of the population? No, it is not, because members of the girls' softball team are likely to have characteristics and opinions different from those of seniors in general. Generally, large random samples are representative of the population. Occasionally, random sampling might give a sample that is not very representative, but it is still a random sample if it was selected using the two criteria.

We obtained random samples using different random mechanisms in the sampling experiments in Sections II and III. Sometimes we used a physical device, such as tossing coins, and sometimes we used a random number table. In each case, the sampling process satisfied the definition for selecting a random sample. Random sampling allows us to construct confidence intervals for the population percentage. If we had not obtained samples randomly, we could not have made the statistical inferences we did.

Deciding If a Sampling Method Gives Random Samples

1. Use a random number table to draw a random sample of five students from your class. Does your sample appear to be representative?

2. For question 1, what proportion of the two-digit random numbers did you use? What proportion did you disregard? Can you think of a more efficient way to assign the two-digit random numbers to students?

3. Which of the following sampling methods produce a random sample from a class of 36 students?

   a. Select the first six students to enter the room.

   b. Select those students whose phone numbers end with the digit 4.

   c. Suppose that the class has 18 boys and 18 girls. Select a sample of 6 students by using a random number table to choose 1 of the 18 boys, then 1 of the 18 girls, then a boy, then a girl, and so on until you have chosen 6 students.

   d. Suppose that the classroom has six rows of chairs with six chairs in each row. Assign the rows the digits 1 through 6. Throw a die and place all the students in the row corresponding to the number on the die in the sample.

   e. Assign each student a number from 1 to 36. The girls get the numbers 1 to 18 and the boys the numbers from 19 to 36. Use a random number table to select six two-digit numbers between 1 and 36, and place the corresponding students in the sample.

4. For each sampling method below, tell which groups in the population are likely to be underrepresented.

   a. To obtain a sample of households, a television rating service dials numbers taken at random from telephone directories.

   b. In 1984, Ann Landers conducted a poll on the marital happiness of women by asking women to write to her.

   c. To determine the percentage of teenage girls with long hair, *Teen* magazine published a mail-in questionnaire. Of the 500 respondents, 85% had hair shoulder length or longer (*USA Today*, July 1, 1985).

    d. To evaluate the reliability of cars owned by its subscribers, *Consumer Reports* magazine publishes a yearly list of automobiles and their frequency-of-repair records. The magazine collects the information by mailing a questionnaire to subscribers and tabulating the results from those who return it.

    e. A college psychology professor needs subjects for a research project to determine which colors average American adults find restful. From the list of all 743 students taking introductory psychology at her school, she selects 25 students using a random number table.

    f. For a survey of student opinions about school athletic programs, a member of the school board obtains a sample of students by listing all students in the school and using a random number table to select 30 of them. Six of the students say that they don't have time to participate, and they are eliminated from the sample.

5. If a sample of 20 adults ends up containing only men, two explanations are possible. The first is that the sampling procedure wasn't random; the second is that the sampling procedure was random but that a nonrepresentative sample resulted. Which explanation would you be more inclined to believe? Explain. (*Hint*: Look at the charts of 90% box plots.)

6. Repeat question 5 but assume that the sample of size 20 had 13 men.

7. Describe how you could actually obtain a random sample of 30 students from the population of students in your school. You may want to consult with someone in the records office.

Other Ways to Obtain a Sample

Obtaining a random sample can be very difficult. For example, there may be no easy way to list all members of a population so that we can assign a number to each member. Even if we could make such a list, it might be very difficult to contact some members in order to include them in the sample.

Suppose we want to conduct a survey to determine which of two candidates will win the next election for dogcatcher in our town. We would like to sample the population of all those who will vote in the election. But who knows who will vote? Nobody. Suppose, then, we define the population as all those who voted in the last election and are still registered. Such information is publicly available, and we could conceivably make a list of all these people, and choose a random sample using the list. But constructing this list would be a lot of work, probably more than we want to do. Alternatively, we could take as the population all voting-age residents of the town and try to take a random sample of them. But who has a complete list of residents of the town? We might find a list of all household addresses, but we would have to know how many people live at each address to produce a complete list for sampling.

Another possibility is to use telephone numbers; more than 90% of households have a telephone. But if we use the telephone book as our list of residents in the population (or at least as a list of households), we will miss all those people with unlisted numbers. Around 20% of all residential phone numbers are not listed in current telephone directories, and this percentage varies depending on the demographic characteristics of the region. Moreover, people with unlisted numbers might have different views about the candidates for dogcatcher than those with listed numbers. If everyone in our town has telephone numbers with the same first three digits (a big if), then we could obtain a sample by dialing these three digits followed by four digits selected at random using a random number table. It will be difficult to reach people who are not home very often, so these people will be less likely to be in the sample. On the other hand, those households with two telephone numbers will be more likely to be in the sample. As you can see, devising a procedure to obtain a random sample for a real question and a real population of interest can be very difficult, if not impossible.

If a method of selecting a sample tends to overrepresent or underrepresent some part of the population, then the method is *biased* (and the resulting samples tend not to be representative of the population). Ideally, pollsters prefer random sampling, but as we saw with the dogcatcher example, random samples are difficult to obtain. In practice, some bias is almost inevitable in the method of sampling.

The rest of this section and Applications 18, 19, and 20 explore the advantages and disadvantages of different sampling methods. You will learn how to evaluate the ways in which a method of selecting a sample might be biased, thereby giving samples that tend not to be representative of the population.

Convenience Sampling

The easiest way to obtain a sample is simply to choose it, without any random mechanism. For example, if we want a sample of 5 from a class of 30 students, we could choose the first 5 students who raise their hands, or choose the 5 in the front row, or choose the 5 tallest, or choose 5 close friends, or choose 5 enemies, or simply name 5 people haphazardly without using any special criteria. Obtaining a sample by such methods is called *convenience sampling*. Convenience sampling uses no explicit random mechanism. It is easy, but is it useful? Can we make confidence interval statements relating the sample to the population, as in Section V, using convenience sampling? Unfortunately, we can't.

Why can't we use convenience sampling to construct confidence intervals? First, our method of constructing a confidence interval (laid out in Sections II through V) depends fundamentally on using a sample selected by random sampling. We made our 90% box plots by observing the variability in different random samples from the same population. Convenience sampling gives us no straightforward way to model the variability from one sample to the next, so we cannot construct box plots or a confidence interval.

Second, convenience sampling is often biased, as you will see in the examples in Applications 18 and 19. With random sampling, we expect no bias. A specific sample may happen not to be representative. However, we know that on average random samples are representative: about 90% of the time the confidence interval will contain the population percentage.

Self-Selected Samples. When people participate in a survey by voluntarily returning a form printed in a newspaper or magazine, they make up a *self-selected sample*, which is one type of convenience sample. People who care enough to respond may not be representative of the whole population. For example, in a mail-in survey of 5,400 *USA Today* readers, an amazing 43% of the respondents in Delaware, Indiana, Kentucky, Michigan, New York, Ohio, and Pennsylvania reported symptoms that pointed to a serious risk for clinical depression. The newspaper notes, however, that "Mail-in surveys always attract the most concerned and motivated. It's not a random sample . . . ." (*USA Today*, July 12, 1985). Such a study cannot reliably tell us the percentage of the overall population at risk for depression.

Judgment Sampling. In another form of convenience sampling, an expert selects a sample that he or she considers representative. For example, a produce buyer might select and taste several grapes from a shipment in order to determine the quality of the grapes as a whole. A judgment sample may give a better estimate than most random samples would, if the expert is really good at selecting the sample. However, there is no easy objective way to quantify such a claim.

In summary, random sampling is useful because we can calculate confidence intervals from samples drawn in this way. However, random sampling is difficult to do in practice. Alternatively, convenience sampling can be easy to do, but it is not always useful for learning about the population as the method may be biased.

Probability Sampling

In the face of all these difficulties in obtaining a sample, what methods do organizations performing large sample surveys actually use? They use procedures called *probability sampling*. Random sampling is one special type of probability sampling.

Probability sampling always includes a random mechanism to choose the members of the sample. Each member of the population is chosen using known probabilities, but the probabilities do not have to be equal; thus, each member of the population is not necessarily equally likely to be chosen. Further, the members of the sample are not necessarily chosen independently.

Three other common types of probability sampling methods are described next. Statisticians have developed formulas for obtaining confidence intervals from probability samples, but the formulas are complicated and we will not discuss them in this book.

Cluster Sampling. Suppose an organization wants to poll voters in a town. It might first select some streets at random in the town, then select some households at random on these streets, and then poll everyone in these households. This sample is not a convenience sample, because at no time does the interviewer decide who to include in the sample. However, it is also not a (simple) random sample, even though each voter in the town has an equal chance of being part of the sample. The reason it is not a random sample is that the people are not chosen independently of one another. If one person is in the sample, every voter in his or her household will be, too; moreover, neighbors on that person's street are more likely to be included than are residents on other streets. This type of sampling is called *cluster sampling*, because the items enter the sample in clusters, not individually.

Stratified Random Sampling. A common type of probability sampling is *stratified random sampling*. In this method, polling organizations divide the population into separate strata, or subgroups, so that each population member is in one, and only one, stratum. Then they take a random sample in each of the strata. For example, to obtain a sample of 40 students from a high school, we could divide the students into the two strata of boys and girls and take a random sample of 20 from each. Alternatively, we could define the four strata as freshmen, sophomores, juniors, and seniors and take 10 students at random from each.

One reason for using stratified random sampling is that the separate strata may be of interest, not just the whole population. In the high school example, we may want to know how the student body as a whole answers the survey question, and we may also want to know how the views of boys and girls compare. Thus, we must make sure we have enough boys and enough girls in the sample, so we define them as the strata.

Another important reason for using stratified random sampling is to insure that the sample is more representative of the population than a (simple) random sample might be. This increased representativeness causes the confidence interval from a stratified random sample generally to be shorter than the confidence interval from a random sample (of the same total size). That is, stratified random sampling usually gives more precise estimates than random sampling.

Systematic Sampling. Another popular type of probability sampling is *systematic sampling*. If you were to select a sample of students from your class by choosing every fifth student who walks into the classroom, you would be using systematic sampling. To use systematic sampling, we first order the members of the population in some way. Next we decide to sample, say, 1 out of every 20. For a 1-in-20 systematic sample, we randomly choose one of the first 20 members of the population and then every 20th member from then on. We might get, for example, members 8, 28, 48, 68, and so on; alternatively, we might get members 17, 37, 57, 77, and so on.

Systematic sampling has several advantages. It is often easier to do than random sampling. It also guarantees that the sample is taken from throughout the ordered population; thus, the sample may be more representative than one from random sampling. A danger, however, is that the way we order the population may be connected to the problem we are studying. For example, suppose we study freeway traffic by taking a systematic 1-in-7 sample of days. We could get Sunday, Sunday, Sunday, which would not allow us to learn much about overall congestion! Systematic sampling is most useful when random sampling is too difficult, and we see no reason for the ordering of the population to create a nonrepresentative sample.

Using Different Sampling Methods

1. Describe how you would select a sample of 30 juniors from your school using the following methods.

   a. random sampling

   b. convenience sampling

   c. sampling by self-selection

   d. stratified random sampling

   e. systematic sampling

   f. cluster sampling

2. Retailers at the local shopping mall want to survey their Saturday customers about their satisfaction with the eating facilities within the mall. One merchant went to business school and learned about the importance of statistics, so he wants to obtain a random sample. He proposes the following method: Interviewers should stand at the center of the mall and select the first 100 people who walk by after 11:00 a.m. He believes this approach will provide a random sample because the interviewers will not exercise any decision over whether or not to include specific individuals in the sample.

   a. What kind of sample would the merchant really get?

   b. In what way might this sampling method be biased?

   c. Describe how the merchant could modify this approach to use a version of systematic sampling.

   d. If the retailer were to use stratified random sampling, what strata would you recommend that he choose?

   e. How would you improve the merchant's sampling procedure?

3. The Educational Testing Service (ETS) needed a representative sample of college students. ETS first divided all colleges into groups of similar ones (such as public colleges with more than 25,000 students, small private schools, and so on). Then they used their judgment to choose one representative school from each group, thus obtaining the sample of schools. Each school in turn picked a sample of students (Freedman, Pisani, and Purves, *Statistics*).

   a. ETS divided the colleges into strata but did not perform stratified random sampling. Explain.

   b. Suggest ways to improve this sampling scheme.

4. Researchers wanted a representative sample of Japanese-Americans living in San Francisco. "The procedure was as follows. After consultation with representative figures in the Japanese community, the four most representative blocks in the Japanese area of the city were chosen; all persons resident in those four blocks were taken for the sample. However, a comparison with Census data shows that the

sample did not include a high-enough proportion of Japanese with college degrees" (Freedman et al.).

    a. What kind of sampling did this study use?

    b. Why do you suppose the sample did not have enough college graduates?

    c. Can you think of a way to improve this sampling scheme? Can you think of a reasonable way to use random sampling to obtain the sample?

5. The headline on page 1 of an Illinois newspaper stated, "More people using drugs at work, survey reports." The article gave the following information: "The survey questioned 227 people who called the national [cocaine] helpline, chosen at random, during a six-week period in February and March . . . Ninety-two percent of the callers said they sometimes worked while under the influence of drugs" (*Rockford Register Star*, March 25, 1985).

    a. What kind of sampling was used?

    b. What population would you say this sample is drawn from?

    c. Describe why this survey does not justify the claim made in the headline.

6. A newspaper article began, "Almost half of the USA's secretaries would rather work for a man than a woman, even though a male boss is more likely to ask them to clean the coffeepot, says a *Working Woman* survey" (*USA Today*, April 23, 1986). This is the result of a "poll of 1,100 readers in the magazine's May issue." Of these readers, 46% prefer to work for a man, 5% for a woman, and 49% say it doesn't matter.

    a. What kind of sampling do you think was used?

    b. What population do the results apply to, according to the newspaper?

    c. In what way might the sampling method be biased? (*Hint*: What kind of secretaries would not read *Working Woman*?)

Analyzing the Largest Sample Survey Ever

In the 1936 presidential election, Franklin D. Roosevelt ran for reelection against Alfred Landon. The *Literary Digest*, a popular magazine that ran preelection polls, had correctly predicted the winner in all presidential races since 1916. In 1936, based on responses from about 2.4 million people, the magazine predicted that Landon would win, 57% to 43%. In fact, Roosevelt won, 62% to 38%. What happened?

To obtain its sample, the magazine compiled a list of about 10 million names from sources such as telephone books, lists of automobile owners, club membership lists, and its own subscription lists. All 10 million people received questionnaires, and about 2.4 million returned them; these people made up the sample.

1. What method of sampling did the magazine use?

2. What percentage of people returned the questionnaire? In what ways do you think people who returned the questionnaire might have differed from those who did not? Do you think that the proportion favoring Roosevelt among those who returned the questionnaire was about the same as the proportion favoring Roosevelt among all those receiving the questionnaire? That is, do you think a **nonresponse bias** existed?

3. Discuss other sources of bias in the magazine's sample selection. In other words, in what ways were the people receiving the *Literary Digest* questionnaire likely to differ from the population of voters in 1936?

4. Why do you think the *Literary Digest* survey successfully predicted the winner from 1916 to 1932 but not in 1936?

It so happened that in 1936 a young man named George Gallup was setting up an organization to do surveys. He predicted the *Literary Digest*'s predictions (with 1% error) well before the magazine published them. Gallup obtained his sample by randomly choosing 3,000 people from the same lists the *Digest* used and mailing them postcards asking how they planned to vote. Gallup also ran a different, larger survey that predicted Roosevelt would win. (For further discussion of this and other election examples, see Freedman, Pisani, and Purves, *Statistics*.)

Good Housekeeping magazine runs an annual "Most Admired Men" poll in which the editors list several columns of prominent men and ask readers to send in their votes. Here is a newspaper comment on this poll. "*Good Housekeeping* magazine has its fifth annual 'Most Admired Men' poll underway, by its very size—circulation in the millions—out-polling all the Nielsens, Trendexes, etc., surveys now cluttering up the nation's opinions. And it's a fascinating project, certainly democratic across the range of G. H. readers, not just the inefficient 1,200 questionees of the noisier nose-countings; some contact 500 persons or less and accept their aggregate word as national opinions, or trends. . . . The *Good Housekeeping* poll strikes us as a far more definitive reflection of opinion" (Newark *Star-Ledger*, August 19, 1985).

   5. Discuss why the *Good Housekeeping* survey might not reflect national opinion as accurately as some other, much smaller polls.

Other Sources of Bias

Even if a survey organization uses random sampling or probability sampling to choose a sample, survey results can be biased for other reasons. For example, people may refuse to respond, they may not tell the truth, the survey may occur at a bad time, it may be poorly worded, or interviewers may not be well trained. Unfortunately, it is always difficult and sometimes impossible to estimate the errors caused by such factors.

Nonresponse Bias

Many people neglect or refuse to respond to surveys; the nonresponse rate can vary from a very low percentage for some governmental surveys to over 90% for some long questionnaires delivered by mail. For example, the city of Laramie, Wyoming, sent out 2,000 surveys to a random sample of Laramie households as part of its second annual Citizen's Attitude survey. Only 481 surveys were returned. What was this survey's response rate? A related problem is that some people, such as those who work long hours, are difficult to contact. Unfortunately, people who refuse to respond and people who are difficult to contact tend to give answers different from other people's.

Untruthful Answers

People give untruthful answers for several reasons. If an interviewer asks a sensitive question, people may give an answer they think is socially acceptable, or tell the interviewer what they think he or she wants to hear. For this reason, in Gallup's presidential election polls, interviewers hand interviewees a ballot that they can mark secretly.

Another problem is that people, not wanting to appear ignorant, will try to answer a question even if they know nothing about the subject. "In one study, educators were asked how they would rank Princeton's undergraduate business program. In every case, it was rated among the top 10 departments in the country, even though Princeton does not offer an undergraduate business major" (*Los Angeles Times*, November 21, 1982).

People often do not remember numbers they are asked about. For example, one study (*Sociological Methods and Research*, November 1981) asked students to report their grade point averages (GPAs). Researchers then determined the actual GPAs. Over 17% of the students reported a GPA 0.4 or more above their actual average, but about 2% of the students reported a GPA more than 0.4 below their actual GPA!

Survey Details

Factors such as the timing of a survey, the specific way questions are worded, and the competence of the interviewers can all affect the survey results. For example, in a National Football League poll in January 1971, football emerged as the nation's favorite sport (Moore, *Statistics: Concepts and Controversies*, first ed.). What happens in January that could have biased this result?

Subtle differences in the phrasing of a question can sometimes cause a large difference in the results. For example, Americans are much more willing to "not allow" speeches against democracy than they are to "forbid" them (Schuman and Presser, *Questions and Answers in Attitude Surveys*). In a 1981 survey, fewer than 10% of the respondents said they would cut programs involving "aid to the needy." But rephrasing the question led 39% to say they supported cuts to "public welfare programs" (*Los Angeles Times*, April 20, 1982).

Similarly, interviewers can sometimes phrase questions to make people respond a certain way. For example, try to say "no" to this question: "Do you favor paying hard-working teachers a little more so that our fine young people can have a decent education?"

Interviewers must not misinterpret people's answers. Consequently, the Census Bureau and other large survey organizations require that their interviewers follow very explicit procedures, and they monitor the interviews closely, with random follow-up by supervisors.

Look at the reproduction on page 49 of part of the questionnaire used by National Crime Survey interviewers. This page is designed to determine background information before getting to the questions on crime. These questions help the Department of Justice to study why people may or may not become victims of crime. Notice that the questionnaire spells out everything the interviewer must say and provides a place to record every response. Such uniformity is necessary to insure that all people surveyed answer the same questions. Otherwise it would be impossible to aggregate the answers into a valid national summary.

Notice also that the questionnaire does not ask, "Are you employed?" Instead, the questions focus on specific activities during the last week in order to get more detailed and reliable information about employment.

PERSONAL CHARACTERISTICS

18. NAME (of household respondent)	19. TYPE OF INTERVIEW PGM 4	20. LINE NO. (cc 12)	21. RELATIONSHIP TO REFERENCE PERSON (cc 13b)	22. AGE LAST BIRTH-DAY (cc 17)	23. MARITAL STATUS (cc 18)	24. SEX (cc 19)	25. ARMED FORCES MEMBER (cc 20)	26. Educa-tion — highest grade (cc 21)	27. Educa-tion — complete that year? (cc 22)	28. RACE (cc 23)	29. ORIGIN (cc 24)
Last / First	**085** 1 ☐ Per. — Self-respondent / 2 ☐ Tel. — Self-respondent / 3 ☐ Per. — Proxy ⎫ Fill 14 on / 4 ☐ Tel. — Proxy ⎬ cover page / 5 ☐ NI — Fill 20–29 and 15 on cover page	**086** Line No.	**087** 1 ☐ Ref. person / 2 ☐ Husband / 3 ☐ Wife / 4 ☐ Own child / 5 ☐ Parent / 6 ☐ Bro./Sis. / 7 ☐ Other relative / 8 ☐ Non-relative	**088** Age	**089** 1 ☐ M. / 2 ☐ Wd. / 3 ☐ D. / 4 ☐ Sep. / 5 ☐ NM	**090** 6 ☐ M / 7 ☐ F	**091** 1 ☐ Yes / 2 ☐ No	**092** Grade	**093** 6 ☐ Yes / 7 ☐ No	**094** 1 ☐ White / 2 ☐ Black / 3 ☐ American Indian, Aleut, Eskimo / 4 ☐ Asian, Pacific Islander / 5 ☐ Other — Specify ↗	**095** Origin

▶ **INTERVIEWER:** If respondent 12–15 go to Check Item A. If 16+ read ↗, then go to Check Item A.

Before we get to the crime questions, I have a few (additional) items that are useful in studying why people may or may not become victims of crime.

CHECK ITEM A
Look at item 3 on cover page. Is box 1 marked?

☐ No — **Ask 30**

Yes — Is this person a new household member?

PGM 5 **100**   1 ☐ Yes — **Ask 30**
   2 ☐ No — **SKIP to Check Item C**

30. How long have you lived at this address?

101 _____ Months (If more than 11 months, leave blank and enter 1 year below.)

OR

102 _____ Years (Round to nearest whole year)

CHECK ITEM B
Is entry in 30 —
☐ 5 years or more? — **SKIP to Check Item C**
☐ Less than 5 years? — **Ask 31**

31. Altogether, how many times have you moved in the last 5 years, that is, since_____ , 197____ ?
(Mo. of Int.)   (5 yrs. ago)

103 _____ Number of times

CHECK ITEM C
Is this person 16 years old or older?
☐ Yes — **Ask 32a**
☐ No — **SKIP to 37a**

32a. What were you doing most of LAST WEEK — (working, keeping house, going to school) or something else?

104 1 ☐ Working — **SKIP to 32c** 6 ☐ Unable to work — **SKIP to 35**
2 ☐ With a job but not at work 7 ☐ Retired
3 ☐ Looking for work 8 ☐ Armed Forces — **SKIP to 36a**
4 ☐ Keeping house 9 ☐ Other — Specify ↗
5 ☐ Going to school

b. Did you do any work at all LAST WEEK, not counting work around the house? (Note: If farm or business operator in HHLD, ask about unpaid work.)

105 1 ☐ Yes
2 ☐ No — **SKIP to 33a**

c. How many hours did you work LAST WEEK at all jobs?

106 _____ Hours — **SKIP to 36a**

33a. If "with a job but not at work" in 32a, SKIP to 33b.
Did you have a job or business from which you were temporarily absent or on layoff LAST WEEK?

107 1 ☐ Yes
2 ☐ No — **SKIP to 34a**

b. Why were you absent from work LAST WEEK?

108 1 ☐ Layoff — **SKIP to 34c**
2 ☐ New job to begin within 30 days — **SKIP to 34c**
3 ☐ Other — Specify ↗   ⎱ **SKIP to 36a**

34a. If "looking for work" in 32a, SKIP to 34b
Have you been looking for work during the past 4 weeks?

109 1 ☐ Yes
2 ☐ No — **SKIP to 35**

b. What have you been doing in the last 4 weeks to find work? Anything else?
Mark all methods used. Do not read list.
Checked with —

110 1 ☐ Public employment agency
* 2 ☐ Private employment agency
3 ☐ Employer directly
4 ☐ Friends or relatives
5 ☐ Placed or answered ads
6 ☐ Other — Specify (e.g., CETA, union or professional register, etc.) ↗

7 ☐ Nothing — **SKIP to 35**

c. Is there any reason why you could not take a job LAST WEEK?

111 1 ☐ No
Yes — 2 ☐ Already had a job
3 ☐ Temporary illness
4 ☐ Going to school
5 ☐ Other — Specify ↗

35. If "layoff" in 33b, SKIP to 36a
When did you last work at a full-time job or business lasting 2 consecutive weeks or more?

112 1 ☐ 6 months ago or less
2 ☐ More than 6 months but less than 5 years
3 ☐ 5 or more years ago
4 ☐ Never worked full time 2 weeks or more ⎱ **SKIP to 37a**
5 ☐ Never worked at all

36a. For whom did you (last) work? (Name of company, business, organization or other employer)

b. What kind of business or industry is this? (e.g., TV and radio mfg., retail shoe store State Labor Department, farm)

113 ☐☐☐

c. What kind of work were you doing? (e.g., electrical engineer, stock clerk, typist, farmer, Armed Forces)

114 ☐☐☐

d. What were your most important activities or duties? (e.g., typing, keeping account books, selling cars, finishing concrete, Armed Forces)

e. Were you —

115 1 ☐ An employee of a PRIVATE company, business, or individual for wages, salary, or commissions?
2 ☐ A GOVERNMENT employee (Federal, State, county, or local)?
SELF-EMPLOYED in OWN business, professional practice, or farm? If yes ↗
Is the business incorporated?
3 ☐ Yes
4 ☐ No (or farm)
5 ☐ Working WITHOUT PAY in family business or farm?

Evaluating Bias in Surveys

Identify any sources of bias in each of the following surveys.

1. The rating service Arbitron estimates the popularity of radio stations in the Los Angeles area. Four times a year, Arbitron takes a random sample of about 10,000 listeners. Every member of the household over age 12 is asked to fill out a diary, showing what he or she listens to every quarter hour from 6:00 a.m. to midnight, for one week. Each diarist receives 50 cents for his or her trouble. At the end of 12 weeks, Arbitron tallies the results from the usable diaries—usually between 33% and 50% of the 10,000 sent out (*Los Angeles Times*, January 31, 1984).

2. One year after the Detroit race riots of 1967, interviewers asked a sample of black residents in Detroit if they felt they could trust most white people, some white people, or none at all. When the interviewer was white, 35% answered "most"; when the interviewer was black, 7% answered "most" (Moore, *Statistics: Concepts and Controversies*).

3. In response to recent proposals for improving the quality of education, a Louis Harris poll was commissioned to find out how teachers feel about certain questions. "We undertook the Metropolitan Life survey of teachers, interviewing a cross-section of elementary and secondary teachers across the United States. In all, we surveyed 1,981 teachers. It can be said theoretically that every public school teacher had an equal chance of being drawn into the final sample." Among the many results: "While they have reservations about merit pay as such, a 71% to 28% majority believe such a system could work if there were an objective standard on which a teacher's individual merit could be judged" (Newark *Star-Ledger*, July 22, 1984).

   The management-oriented Educational Research Service also conducted a survey of teachers at about the same time. Among other results, this study found that 50.8% of a random sample of 1,013 teachers "either agreed or tended to agree that merit or incentive pay should be given to teachers who meet appropriate performance criteria" (Newark *Star-Ledger*, September 22, 1984). Comment on the apparent difference in opinion on merit pay between the surveys.

4. To find out how people reacted to the clothes of vice-presidential candidate Geraldine A. Ferraro, researchers ran a survey shortly after the 1984 Democratic convention in three locations: the Wall Street area of New York City, State Street in Chicago, and Crown Center in downtown Kansas City. The researchers stopped people at random and asked them if they had seen the Democratic convention on television. Those who had were not used. Those who had not "were asked if they would be willing to contribute a minute or two of their time to help a woman candidate choose a suitable picture for a campaign poster. We wanted to enlist only those who had a positive attitude toward women running for office." The 347 respondents were then shown pictures of women wearing three outfits, and the pictures

did not show the women's faces. Then the respondents were asked several questions (*Los Angeles Times*, August 3, 1984).

5. The following quotation is from a report on a survey of high school students' views of nuclear war. "[It] is based on 5,553 responding high school students (10th, 11th, and 12th graders). Thirty-three northern New Jersey public and private high schools, selected solely upon their willingness to have the questionnaire administered to a group of their students, participated in the study. The students came from various economic backgrounds and environments: inner city (297), urban and suburban middle class (2,217), affluent suburbs (2,313), and rural areas (722). They ranged from fewer than 50 in social studies or other classes in some schools, to virtually the entire 10th, 11th, and 12th grades in others." Among other results reported, one question concerned the likelihood of a nuclear blast caused by an act of war somewhere on earth in the next 20 years. Thirty-seven percent said this event is likely, 37% said it is somewhat likely, and 26% said it is unlikely (*Physicians for Social Responsibility*, July 1984).

6. On the recent deregulation of banking, "[the head of California's Security Pacific Bank] reckons the higher interest accounts, and all the other new financial services, are designed for the most affluent 15% to 20% of Security Pacific Bank's customers. By extension—as 2 million customers are surely a sample of the general population—the new world of deregulated finance benefits the top-earning 15% to 20% of U.S. households" (*Los Angeles Times*, December 4, 1983).

7. A Gallup poll found that 81% of U.S. parents say they have spoken with their teenagers about the dangers of drinking and driving. Only 64% of the teens say they remember such a discussion (*USA Today*, December 19, 1984).

8. The U.S. census of 1980 states that 32,194 Americans are 100 years old or older. However, Social Security figures show only 15,258 adults of this advanced age (*Los Angeles Times*, December 16, 1982).

9. In a 1983 survey of fourth graders (nine-year-olds), *Weekly Reader* found that 30% felt peer pressure from other children to drink alcoholic beverages (*Cincinnati Enquirer*, April 22, 1986). The newspaper article did not publish the wording of the question. (*Hint*: Try to write a question that you are sure nine-year-olds will understand that asks if they feel peer pressure to drink.)

10. In a census in Russia, 1.4 million more women than men reported that they were married (*U.S. News & World Report*, August 30, 1976).

VII. LARGE SURVEYS

Large surveys, such as the Gallup poll, differ from the surveys we studied in Sections II through V in four major ways:

1. As we saw on page 2, large survey organizations report an "error attributable to sampling and other random effects," or "sampling error," rather than confidence intervals.

2. Large surveys use a 95% box plot rather than the 90% box plot we have used.

3. The sample size is at least several hundred and is usually about 1,500.

4. As we discussed in Section VI, the samples for large surveys are usually not obtained using random sampling. Instead, pollsters use a form of probability sampling for which they can compute confidence intervals.

You will learn more about these differences in this section.

Reminder

The 90% confidence interval contains all of the population percentages for which the sample proportion is likely.

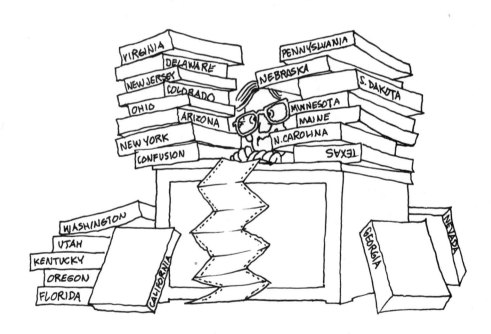

Application 21

Calculating Sampling Error

According to a study reported in *USA Today* (see page 23), about 60 of the 100 divorced couples with children (sample proportion 0.60) are "unfriendly." From the box plots for samples of size 100 (page 95), we find that the 90% confidence interval for the percentage of divorced couples with children who are unfriendly is 55% to 65%. Another way to say the same thing is that we think that about 60% of all divorced couples with children are unfriendly, with a *sampling error* of 5% either way.

The term sampling error refers to the size of error that occurs because the sample proportion from a random sample is not usually the same as the population percentage. (Less frequently used synonyms for sampling error are margin of error, chance error, sampling tolerance, standard margin of error, and error attributable to sampling.) The sampling error *does not* include errors resulting from possible sources of bias, such as nonresponse bias, untruthful replies, or bad wording of questions.

In each of the surveys described below, find

    a. the sample proportion.

    b. the 90% confidence interval.

    c. the sampling error.

1. In a survey of 20 people, 10 people are TV addicts.

2. A survey of 40 women finds 12 who do not work.

3. A survey of 40 students finds 20 who like cafeteria food.

4. A random sample of 100 boys includes 35 with conduct disorders.

5. A random sample of 100 students includes 60 nonsmokers.

Complete these sentences by looking at your answers to the questions above.

6. For a sample of size 20, the sampling error is about _____%.

7. For a sample of size 40, the sampling error is about _____%.

8. For a sample of size 100, the sampling error is about _____%.

9. As the sample size increases, the sampling error _____ .

95% Box Plots

Let's review confidence intervals. If we take random samples from a population and construct a confidence interval from each sample, then the population percentage should be inside 90% of the intervals we construct. In other words, we expect that in 10 out of every 100 surveys, the confidence interval will *not* contain the population percentage.

If large polling organizations used 90% box plots, they would be wrong in about 10 of every 100 surveys. Thus, they use 95% box plots and are wrong in about 5 of every 100 surveys. Using 95% box plots lengthens both the box and the confidence interval.

Large Sample Sizes

Perhaps you suspect that polling organizations do not compute confidence intervals from charts of 95% box plots. If so, you are right! Now that you understand the basic ideas of confidence intervals, we can simplify our procedure.

A simple formula is available to determine the sampling error in a survey that uses random sampling. It is

$$2\sqrt{\frac{p(1-p)}{n}} \, ,$$

where p is the sample proportion and n is the sample size. The formula gives a decimal, which we can convert to the corresponding percentage. To use this formula, you need $np \geqslant 5$ and $n(1-p) \geqslant 5$. The formula can be derived using a complicated statistical theory that we will not go into. This theory says that when we use this formula, the population percentage will be in the confidence interval at least 95% of the time.

For example, suppose we take a sample of size 100 and get 70 *yes*es. Then

$$n = 100 \, ,$$

$$p = \frac{70}{100} = 0.70 \, ,$$

and the sampling error is

$$2\sqrt{\frac{p(1-p)}{n}} = 2\sqrt{\frac{0.70\,(1-0.70)}{100}}$$

$$\cong 0.09 \text{ or } 9\% \, .$$

The 95% confidence interval for the true population percentage is 61% to 79%.

For the remainder of this book, round any sampling error to the nearest whole percent.

Finding the Sampling Error

Assume random sampling and find the sampling error as a percentage for each of the surveys below. Use the formula $2\sqrt{p(1-p)/n}$.

1. In a sample of 25 students, 20 said *yes*.

2. In a sample of 100 students, 80 said *yes*.

3. In a sample of 400 students, 320 said *yes*.

4. In a sample of 1600 students, 1280 said *yes*.

5. Look over your answers to questions 1 through 4. When you multiply the sample size by 4, what happens to the sampling error?

6. In a sample of 200 taxpayers, 76 cheated on their income tax (*Los Angeles Times*, April 13, 1986).

    a. Find the sampling error.

    b. Find the 95% confidence interval for the percentage of taxpayers who cheat.

7. In a study of 500 children, ages three to seven, 68% didn't know their home phone number (*USA Today*, August 7, 1985).

    a. Find the sampling error.

    b. Find the 95% confidence interval for the percentage of three- to seven-year-olds who don't know their home phone number.

8. If $p = 0.50$, what sample size gives a sampling error of

    a. 1%?

    b. 2%?

    c. 3%?

    d. 4%?

    e. 5%?

    f. 6%?

    g. 7%?

    h. 8%?

Working with Large Surveys

In each of the surveys described below, assume the sample is obtained by random sampling and

a. identify or compute the sample proportion.

b. find the sampling error to the nearest percent using the $2\sqrt{p(1-p)/n}$ formula.

c. find the 95% confidence interval.

1. The National Assessment of Educational Progress asked 2,000 17-year-olds to estimate the answer to 3.04 × 5.3. The percentage who got the right answer was 36.6 (*The Third National Mathematics Assessment: Results, Trends, and Issues*, 1983).

2. When the Gallup survey interviewed 416 teenagers, ages 13 to 18, on their drinking and drug use, 108 said they used marijuana (*Los Angeles Times*, September 8, 1984).

3. Of 1,557 cars traveling on a residential street, 701 made a full stop at a new stop sign (*Los Angeles Times*, September 6, 1984).

4. A survey by A. C. Nielson Co. of 1,100 owners of videocassette recorders found that 65% of those questioned used the "fast forward" feature to skip commercials on TV programs they had taped (*Los Angeles Times*, September 2, 1984).

5. In Puerto Rico, investigators found that 7 of 61 infants hospitalized with salmonella infections had probably acquired their infection from pet turtles (*Harvard Medical School Health Letter*, September 1984).

6. A survey of 300 mathematically gifted teenagers found that 70% were nearsighted (*Los Angeles Times*, January 6, 1984).

7. In a survey of 506 full-time college students, 48% said they drink at least once a week (*USA Today*, April 1, 1985).

8. Of 2,265 high school seniors surveyed, 73% said they had used crib notes to cheat on a test (*Los Angeles Times*, April 17, 1986).

9. In 200,000 taste tests, the Coca-Cola Company found that 55% of the participants preferred the "New Coke" to the old Coke (*Los Angeles Times*, July 30, 1985).

10. In a *USA Today* survey of 999 college students, 88% said that they want to have children (May 13, 1986).

We have learned that survey organizations use probability sampling, not true random sampling. Therefore, the sampling error reported in a newspaper is not necessarily the same one we would compute using the $2\sqrt{p(1-p)/n}$ formula. Depending on the type of probability sampling, the sampling error could be larger, about the same, or even smaller than the sampling error obtained from random sampling. Generally, it will be a bit larger.

In the surveys described below,

    a. compute the sampling error to the nearest percent using the $2\sqrt{p(1-p)/n}$ formula, as if the sample were random.

    b. decide if the reported sampling error is the same, larger, or smaller than that for a random sample.

    c. find the 95% confidence interval using the reported sampling error.

11. A survey of 1,574 registered voters found that 60% would vote for Ronald Reagan for president. The pollster reported the sampling error as 4 points (*Los Angeles Times*, September 2, 1984).

12. *Los Angeles Times* reporters randomly surveyed 113 businesses to determine why no freeway congestion occurred during the Summer Olympics in Los Angeles. Of those business managers surveyed, 24% said they had continued to work a five-day week but had altered their hours. The article said that the error attributable to sampling for such a survey is 9% (September 2, 1984).

13. When the *Los Angeles Times* surveyed 1,093 regular freeway commuters, 14% reported leaving town during at least part of the Olympic Games. The article says that "such a poll has a margin of error of about 3%" (September 2, 1984).

14. Pollsters questioned 874 men, of whom 33% said working mothers are worse mothers than those who stay at home. The reported margin of error was 3 points (*Los Angeles Times*, September 10, 1984).

15. In a telephone poll of 1,504 adults, 58% reported attending an art exhibition in 1984. A *New York Times* article about the poll says that "the results are within plus or minus three percentage points of what they would have been had the entire population been polled" (December 4, 1984).

16. The following letter appeared in the *Los Angeles Times* on September 7, 1984.

Polls

Polls, polls, polls, that's all you hear every week during presidential campaigns! I say they're hog-wash.

No pollster has ever called me or knocked on my door to get my opinion on any issue whatsoever, and until such time as one ever singles me out for my opinion, I will continue to disbelieve all polls, and denounce them all as nothing but propaganda devices.

Write a response to the letter's author.

National Surveys

Polling organizations conduct most national surveys either by telephone interviews or by personal interviews. Using a mail questionnaire is another possibility, but the nonresponse rate tends to be higher, leading to less reliable results.

Large survey organizations occasionally print explanations of how they do their polling. Here are reports from the *New York Times*/CBS News poll and the Gallup survey. You will now be familiar with many of the technical words and concepts in these articles. You will also find some terms that we have not discussed and that will probably be unfamiliar to you. Note that for sampling error, Gallup uses the term *sampling tolerance*.

How the Poll Was Conducted

The latest *New York Times*/CBS News Poll is based on telephone interviews conducted May 29 through June 2 with 1,509 adults around the United States, excluding Alaska and Hawaii.

The sample of telephone exchanges called was selected by a computer from a complete list of exchanges in the country. The exchanges were chosen to insure that each region of the country was represented in proportion to its population. For each exchange, the telephone numbers were formed by random digits, thus permitting access to both listed and unlisted residential numbers.

The results have been weighted to take account of household size and to adjust for variations in the sample relating to region, race, sex, age and education.

In theory, in 19 cases out of 20 the results based on such samples will differ by no more than 3 percentage points in either direction from what would have been obtained by interviewing all adult Americans. The error for smaller subgroups is larger. For example, the margin of sampling error for Democrats or Republicans is plus or minus 4 percentage points.

In addition to sampling error, the practical difficulties of conducting any survey of public opinion may introduce other sources of error into the poll.

Source: *New York Times*, June 5, 1985.

Design of the Gallup Sample

The Gallup Organization, Inc., maintains a national probability sample of interviewing areas for use in personal interview surveys. The sampling procedures used in the selection of these areas, and in the selection of households and individuals within these areas, are designed to produce sample estimates of the adult population (18 years of age or older) living in the United States (the 50 states and the District of Columbia), *excluding* military personnel living on military bases and persons residing in institutions, such as prisons or hospitals.

The sample follows a replicated, multi-stage area probability design, using stratification by geography, urbanization, and size of community. The areal selection is to the block level in urban areas and to segments of townships (or equivalents) in the case of rural areas. Approximately 300 sampling locations are used in full-scale national surveys.

The sample design first stratified the population by size of community and urbanization, using 1980 census data, into the following categories:

1. Population of central cities of 1,000,000 of more persons.
2. Population of central cities of 250,000 to 999,999 persons.
3. Population of central cities of 50,000 to 249,999 persons.
4. All population not covered in 1, 2, or 3 above, yet located in urbanized areas (as defined by the Bureau of Census).
5. Population of cities and towns (incorporated places and census-designated places) of 2,500 to 49,999 persons.
6. Population of towns and villages (incorporated places and census-designated places) of less than 2,500 persons.
7. All other population.

Population in each of these city size/urbanization categories was further stratified into eight geographic regions: New England, Middle Atlantic, East Central, West Central, Southeast, South Central, Mountain, and Pacific. Within each community size/urbanized area/regional stratum, the population was then arrayed in a serpentine geographic order and zoned into equal-sized units. Replicate sets of localities were selected in each zone, with probability of selection of each locality proportional to its population size.

In the next stage of sample selection, the designated localities were further subdivided and subdivisions were drawn with the probability of selection proportional to the size of population in each subarea. In localities for which subdivision population data are not reported, small definable geographic areas were selected with equal probability.

For each personal interview survey, within each subdivision for which block statistics are available, a separate sample of blocks or block groups is drawn, with probability of selection proportional to the number of dwelling units. In all other subdivisions or areas, blocks or segments are drawn with equal probability.

Within each cluster of blocks and each segment, a randomly selected starting point is designated on a map of the area. Starting at this point, the interviewer is required to follow a predetermined travel path, attempting an interview at each household on this route. The interviewer continues until his or her assignment, which includes a set number of interviews with male and female respondents, has been completed.

(continued)

Interviewing is conducted on weekends or on weekday evenings, when adults are most likely to be at home. Only one interview is conducted in each household.

Allowance for persons not at home is made by a "times-at-home" weighting procedure rather than by call-backs. This method helps reduce the sample bias that would otherwise result from underrepresentation of persons who are seldom at home. . . .

While an estimate of the standard error for any obtained proportion can be computed, it may be helpful to consider the "typical" range of sampling error found in Gallup surveys. Based on numerous estimates, the sampling tolerance for mid-range proportions obtained using a standard 1,500-case national sample is approximately plus or minus 3 percentage points, at the 95% confidence level. For proportions outside the middle range (e.g., reflecting 90%—10% or 80%—20% divisions of behavior or opinion), somewhat smaller sampling tolerances are appropriate. For proportions based on sub-samples (e.g., men only) larger tolerances are appropriate.

It should be noted that these tolerances reflect random variations in the sampling process, design effects due to clustering and weighting, and other random variations introduced in interviewing and data processing. The tolerances *do not* take into account sources of nonrandom error or other possible biases. While every effort is made to avoid such errors, it should be borne in mind that sampling tolerances alone do not reflect all possible sources of inaccuracy in the survey research process.

Source: The Gallup Organization, Inc.

Application 24

Comparing the New York Times/CBS and Gallup Polls

1. Which poll uses the telephone and which uses personal interviews?

2. What is the sample size for each poll?

3. What sampling error does each poll report?

4. When a person is not at home, does Gallup call back later? How does Gallup try to minimize the probability that the person is not at home?

5. Does the article say how the *New York Times*/CBS poll deals with the problem of nonresponse?

6. Does either poll use random sampling? If not, what kind of sampling does each use?

7. What variable(s) does the Gallup poll use for stratification? What variables does the *New York Times*/CBS poll use?

8. Assuming that 170 million adults live in the United States, what is the probability that an adult would be interviewed in a specific Gallup survey?

9. Does either article mention possible sources of error other than sampling error? Does either article give examples of such sources of error? Does either give any number for the potential size of such error?

Planning and Carrying Out Your Own Survey

Work on your own or in a small group on this project.

1. Write a *yes-no* question on a topic that interests you for a survey of 40 students.

2. Decide exactly what population you will sample from. For example, your population could be all girls in your school, all seniors, or all students enrolled in history courses.

3. As a pretest, ask your question to a few members of your class to see if they interpret it exactly as you intend. How can you increase the chances that students will tell the truth? Will you ask for a verbal response? A secret ballot answer? Change your procedures or the wording of your question, if necessary.

4. How will you use random sampling to select 40 students?

5. Obtain 40 students for your sample and ask your question.

6. What is your sample proportion?

7. What is the 95% confidence interval for the percentage of *yes*es in the population? (Use the formula.)

8. Write an article for the school newspaper reporting the results of your survey. Explain the meaning of the confidence interval in your article.

Assessing Opinions About Populations

A Vanderbilt University psychologist had each of 12 pairs of siblings wear one of his or her T-shirts to bed three nights in a row. The T-shirts were then put in individual boxes with small openings in the lids. Each child received two boxes, one of which contained the T-shirt of his or her brother or sister. Of the 24 children, 19 (sample proportion 0.79) identified, by smell alone, the T-shirt worn by a brother or sister (*Science 83*, March 1983).

In the next application, you will answer questions about similar situations. Study these sample questions and answers first.

a. Construct the 95% confidence interval for the percentage of children that choose the correct box.

The sampling error is

$$2\sqrt{\frac{p(1-p)}{n}} = 2\sqrt{\frac{0.79\,(1-0.79)}{24}}$$

$$= 0.17, \quad \text{or} \quad 17\% .$$

Thus the 95% confidence interval is 79% ± 17%, or 62% to 96%.

b. Do you think that children have some ability to identify siblings' clothing by smell?

Yes, we do. If children did not have this ability, in the long run they would choose the correct box 50% of the time. However, 50% is not in the confidence interval of 62% to 96% for the true percentage of children that choose the correct box.

As another example, suppose a teacher claims that exactly 10% of people who have recently left teaching say that students' lack of motivation was one of the main reasons they left. A study of 500 recent former teachers found that 8% gave this reason (*American Educator*, Summer 1986).

a. Construct the 95% confidence interval for the percentage of recent former teachers who would give lack of student motivation as one of the main reasons they left teaching.

The sampling error is

$$2\sqrt{\frac{(0.08)(0.92)}{500}} = 2\% ,$$

so the 95% confidence interval is 6% to 10%.

b. Should you tell the teacher that he or she is wrong?

No, you should not—for statistical as well as political reasons. The 95% confidence interval for the true percentage of recent former teachers giving lack of student motivation as a reason includes the teacher's figure of 10%. Thus, 10% could well be the exact percentage.

Assessing Opinions

For each of the surveys below, assume the samples are random.

1. A study of 300 mathematically gifted children found that 20% were left-handed (*Los Angeles Times,* January 6, 1984).

    a. Construct the 95% interval for the percentage of mathematically gifted children who are left-handed.

    b. About 8% of the whole student population is left-handed. Do you think the proportion of left-handers is greater among mathematically talented students than among students in general? Explain.

2. Out of the same 300 mathematically gifted children, 60% had allergies or asthma.

    a. Construct the 95% confidence interval for the percentage of mathematically gifted children who have allergies or asthma.

    b. Ten percent of the whole student population has allergies or asthma. Do you think a greater percentage of mathematically gifted children have allergies or asthma than do students in general? Explain.

3. A handwriting analyst examined 10 pairs of handwriting samples. One sample in each pair was from a psychotic and the other was from a normal person. The handwriting analyst correctly identified the psychotic in 6 of the 10 pairs (Larsen and Stroup, *Statistics in the Real World*).

    a. Construct the 95% confidence interval for the percentage of pairs of handwriting samples the analyst will get correct in the long run.

    b. Do you think the analyst can identify the handwriting of a psychotic? Explain.

4. Observations of 255 right-handed mothers during the first four days after delivery showed that 212 of the mothers held their babies on the left (Nemenyi, P. et al., *Statistics from Scratch*).

    a. Construct the 95% confidence interval for the percentage of right-handed mothers who hold their babies on the left.

    b. Do you think it is just as likely for a right-handed mother to hold her baby on the left as on the right? Explain.

5. In a survey of 416 teenagers ages 13 to 18, 23% said they do not drink (*Los Angeles Times*, September 8, 1984).

    a. Construct the 95% confidence interval for the percentage of teenagers who say they do not drink.

b. Among people 18 and older, 35% do not drink. Do you think teenagers aged 13 to 18 tend to drink more than those 18 and older? Explain.

6. Ask a friend to estimate the percentage of college undergraduates who say that they are bored in class.

   a. A Carnegie Foundation survey of 5,000 undergraduates found that 36.9% say they are bored in class. Construct a 95% confidence interval for the percentage of undergraduates who say they are bored in class.

   b. Do you think your friend's estimate is the true population percentage? Explain.

VIII. A CAPTURE-RECAPTURE METHOD

Sometimes the people responsible for managing wildlife populations need to count the total number of animals in a population. For example, they might want to know how many deer are in a forest, how many fish are in a lake, or how many seals are on an island. It is impossible to count these animals directly, so naturalists use ingenious *capture-recapture methods*. These methods include several statistical procedures, some quite complicated. In this section, we will discuss the simplest type of capture-recapture method.

In 1970, naturalists wanted to estimate the number of pickerel fish in Dryden Lake in central New York State. They captured 232 pickerel, put a mark on their fins, and returned the fish to the lake. Several weeks later, another sample of 329 pickerel fish were captured. Of this second sample, 16 had marks on their fins (Chatterjee in Mosteller et al., *Statistics by Example: Finding Models*).

Let N be the total number of pickerel fish in the lake. Because the proportion of marked fish in the population should be approximately equal to the proportion of marked fish in the sample, we can write the following equation.

$$\frac{\text{number of marked pickerel fish in the population}}{\text{total number of pickerel fish in the population } (N)} = \frac{\text{number of marked pickerel fish in the sample}}{\text{number of pickerel fish in the sample}}$$

Then we can estimate N by solving this equation. Thus, in Dryden Lake,

$$\frac{232}{N} = \frac{16}{329}$$

$$16N = (232)(329)$$

$$N = \frac{(232)(329)}{16}$$

$$N = 4{,}770.5$$

The *estimate* for the number of pickerel fish in the lake is 4,771.

Application 27

Practicing with the Capture-Recapture Equation

1. Your teacher has a container with a number of cards in it. Draw out 25 of the cards and mark them with a pen or pencil. Return the cards to the box and mix them thoroughly. (It is difficult to mix them well.) Draw a sample of size 20 and count the number marked. What is the estimate for the number of cards in the box?

2. Suppose that naturalists catch, tag, and release 50 deer in a forest. After allowing time for the tagged deer to mix with the others, they catch a sample of 100 deer, 10 of which have tags. What is the estimate for the number of deer in the forest?

3. Suppose that wildlife workers capture 328 penguins on an island, mark them, and allow them to mix with the rest of the population. Later, they capture 200 penguins, 64 of which are marked. What is the estimate for the number of penguins on the island?

4. Suppose that the high school in a town has 500 students. A random survey of 200 people in the town finds 40 high school students. What is the estimate for the number of people in the town?

5. Visitors conducted a capture-recapture experiment to determine the number of taxicabs in Edinburgh, Scotland. On the first day, observers saw 48 taxicabs. The next day they observed 52 cabs, 10 of which they had seen the previous day (The Wildlife Society, October 1978). What is the estimate for the number of taxicabs in Edinburgh?

6. In a study of raccoons in a certain region of northern Florida, 48 animals were captured using cages baited with fish heads. The raccoons were marked and released. In the following week, 71 raccoons were captured, 31 of which had been marked (Pollock in Brook and Arnold, *The Fascination of Statistics*). What is the estimate for the number of raccoons in this region?

Experimenting with the Capture-Recapture Model and Its Assumptions

The capture-recapture equation and its use can, at first, appear deceptively simple. In this application we do several experiments to learn what the assumptions are behind this method and why they matter.

1. Use the container from question 1 of Application 27, which contains 25 marked cards. Mix the cards thoroughly, draw a sample of size 20, and count the number marked. What is the estimate for the number of cards in the box this time?

2. Repeat question 1 four more times. Write the four new estimates.

The capture-recapture method gives only an estimate for the number of cards in the container. Each time you did the experiment, you probably came up with a different estimate. Perhaps you were even surprised at how different the estimates were from one another. Can you guess what we need in addition to the estimate? We need a confidence interval to give us an idea of how precise the estimate is. Later in this section we will learn how to construct a confidence interval for this capture-recapture method.

First, however, we must be aware of some potential problems with this method and the assumptions it is based on. For example, suppose some of the marked animals become afraid of being caught again and avoid traps. How will this behavior affect the estimate of the population size? The following experiments and questions deal with this situation.

3. Ask your teacher how many cards are in the box.

4. Did your estimates from questions 1 and 2 tend to be too large or too small? (They should have been about evenly divided between too large and too small.)

5. Remove 10 of the *marked* cards from the box. These cards represent animals who are "trap shy" and don't want to be captured again. Repeat questions 1 and 2. Write the five estimates.

6. Did the estimates from question 5 tend to be too large or too small? (The population size is still the same as before.)

7. Complete this sentence: When some of the marked animals hide, the estimate of the population size tends to be too _____ .

8. Sometimes some "trap happy" animals are easier to capture and easier to recapture than others. Thus, an animal captured the first time is also likely to be in the second sample. What do you think this behavior will do to the estimate of the population size? Design an experiment with the box of cards to find out if the estimated population size would tend to be too big or too small.

Questions 5 through 8 show that when we use this capture-recapture method, we are assuming that all animals are equally likely to be captured in both trappings. When this assumption is not true, the method can give bad estimates.

In using this capture-recapture method, we have also made two additional assumptions. We assumed that the marks would not be removed, wear off, or become invisible in some way before the recapture. Finally, we assumed that the population was closed during the period of the study—that is, it had no additions due to births or animals entering the area nor deletions due to deaths or animals leaving the area.

9. Complete this sentence. If some animals lose their marks during the study, the estimate of the population size will tend to be too _____ . (If you are unsure of the answer, design and run an experiment to find out.)

10. Suppose the time between the capture and the recapture is too long and some marked animals die. Suppose also that some new animals are born so that the population size remains constant. Will the deaths tend to make the estimate of the population size too large or too small? Explain.

11. Reread question 5 of Application 27. Discuss which assumptions of the capture-recapture model this example may violate.

To use this capture-recapture method, naturalists and statisticians must be convinced that the three basic assumptions are satisfied reasonably well. More complicated capture-recapture methods are available if these assumptions cannot be satisfied.

Confidence Intervals for Capture-Recapture Problems
Using Charts of 90% Box Plots

As you have seen, the method of capture-recapture gives an estimate of the population size. We want a confidence interval to accompany the estimate. Now we will see how to adapt our method for the sample survey problem to give a confidence interval for this capture-recapture method.

Suppose we capture and tag 150 birds in a park. We later capture 100 birds and see that 30 have tags. Thus, the sample proportion of tagged birds is

$$\frac{30}{100} = 0.30$$

What percentage of all birds in the park are tagged? Our estimate is 30%, but we don't know the population percentage for sure. Checking the charts of 90% box plots for a sample of size 100 on page 95, we see that a sample proportion of 0.30 is a likely result from populations with from 25% to 35% *yes*es. Therefore, it is likely that from 25% to 35% of the birds in the park have tags. The 90% confidence interval for the *percentage* of birds in the population that have tags is 25% to 35%.

We can now use this confidence interval for the percentage of birds tagged to construct a confidence interval for the total *number*, N, of birds in the park. Suppose the true percentage of tagged birds in the population is at the lower end of the confidence interval, 25%. We estimate the total number of birds using this percentage:

$$\frac{\text{number of tagged birds in the population}}{\text{total number of birds in the population } (N)} = \text{smallest percentage in confidence interval}$$

$$\frac{150}{N} = 0.25$$

Solving this equation, $N = 600$.

Similarly, suppose the true percentage of tagged birds is at the upper end of the confidence interval, 35%. Then we estimate the total number of birds using this percentage:

$$\frac{\text{number of tagged birds in the population}}{\text{total number of birds in the population } (N)} = \text{largest percentage in confidence interval}$$

$$\frac{150}{N} = 0.35$$

$$N \cong 429$$

Thus, a 90% confidence interval for the *number* of birds in the park is 429 to 600. Remember that for every 100 times we construct a confidence interval this way, the true number of animals will be inside the confidence interval about 90 times. Remember also that to use this capture-recapture method, the assumptions discussed in Application 28 need to be satisfied.

Finding Confidence Intervals for Capture-Recapture Problems Using Charts of 90% Box Plots

1. Suppose you capture, tag, and release 200 fish in a lake. You later capture a sample of size 20 and find that 6 have tags. Use the chart of 90% box plots on page 92 to answer these questions.

   a. Is catching 6 tagged fish out of 20 a likely result if 15% of the fish in the lake are tagged?

   b. Is this result likely if 35% of the fish in the lake are tagged?

   c. List the population percentages that are likely.

   d. What is the 90% confidence interval for the *percentage* of tagged fish in the lake?

   e. What is the 90% confidence interval for the *number* of fish in the lake?

2. Suppose biologists capture, tag, and release 100 snakes in a desert. They then capture a sample of size 100, 40 of which have tags. Use the chart of 90% box plots on page 95 to answer these questions.

   a. Find the 90% confidence interval for the percentage of tagged snakes in the desert.

   b. Find the 90% confidence interval for the number of snakes in the desert.

3. Suppose a biology class captures, marks, and releases 75 mice in a field. Later they capture a sample of 80 mice, 40 of which have marks. Use the chart of 90% box plots on page 94 to answer these questions.

   a. What is the 90% confidence interval for the percentage of marked mice in the field?

   b. What is the 90% confidence interval for the number of mice in the field?

4. Suppose visitors note 100 taxicabs in a city. The next day they observe 100 taxicabs, and 35 are ones they saw the day before.

   a. Find the 90% confidence interval for the percentage of taxicabs the visitors originally noted.

   b. Find the 90% confidence interval for the number of taxicabs in the city.

Using the $2\sqrt{p(1-p)/n}$ Formula to Construct Confidence Intervals for Capture-Recapture

Gorbatch seal rookery, a breeding ground on St. Paul Island in Alaska, wanted to estimate the number of fur seal pups in the rookery. In early August 1961, wildlife workers captured and marked 4,965 pups by shaving some of the black hair from the tops of their heads. They then released the pups and allowed them to mix with the others. In late August, when the workers captured a sample of 900 pups, 218 of them had marks (Chatterjee in Mosteller et al., *Statistics by Example: Exploring Data*).

We want to estimate the number of pups in the rookery and to construct a confidence interval for this number. However, we do not have charts of 90% box plots to use for a sample of size 900. Instead, we can use the $2\sqrt{p(1-p)/n}$ formula to obtain a 95% confidence interval. (Recall that this formula gives a 95% confidence interval, not a 90% one.)

The sample proportion is

$$\frac{218}{900} = 0.24 .$$

The true percentage of tagged pups in the population is probably not exactly 24%, so we will construct a confidence interval to get limits of likely population percentages. Because a sample of size 900 with $p = 0.24$ has a sampling error of

$$2\sqrt{\frac{(0.24)(1-0.24)}{900}} = 0.03 \text{ or } 3\% ,$$

the 95% confidence interval for the percentage of marked pups in the population is 21% to 27%. In other words, if we choose a random sample of size 900 and find a proportion 0.24 marked in our sample, this sample proportion is a likely result from populations with from 21% to 27% marked.

Next we use this confidence interval for the capture-recapture problem exactly as we did when we obtained the confidence interval from the charts of 90% box plots. If the true percentage of marked pups in the population is at the lower end of the confidence interval, 21%, we use

$$\frac{\text{number of marked pups in the rookery}}{\text{total number of pups in the rookery } (N)} = \text{smallest percentage in confidence interval} ,$$

$$\frac{4,965}{N} = 0.21 .$$

Solving gives $N = 23,643$. Similarly, using the largest percentage in the confidence interval gives

$$\frac{4,965}{N} = 0.27 ,$$

or $N = 18,389$. Thus, the 95% confidence interval for the number of pups is 18,389 to 23,643. These population sizes are likely to have 24% marked in a sample of size 900.

Finding Confidence Intervals for Capture-Recapture Problems Using the Formula

For these questions, use the $2\sqrt{p(1-p)/n}$ formula when you must find sampling errors and confidence intervals.

1. Suppose we capture, tag, and release 100 fish in a lake. We then catch a sample of size 100, 40 of which have tags.

   a. What is the estimate for the percentage of tagged fish in the lake?

   b. What is the sampling error for the percentage of tagged fish in the lake?

   c. What is the 95% confidence interval for the percentage of tagged fish in the lake?

   d. What is the 95% confidence interval for the total number of fish in the lake?

2. Suppose rangers catch, tag, and release 180 deer in a game preserve. They then capture a sample of size 90, 15 of which have tags.

   a. What is the estimate for the percentage of tagged deer in the preserve?

   b. What is the sampling error for this percentage?

   c. What is the 95% confidence interval for the percentage of tagged deer in the preserve?

   d. What is the 95% confidence interval for the number of deer in the preserve?

3. Suppose the number of students in the town high school is 500. A random survey of 200 people in the town finds 40 high school students. What is the 95% confidence interval for the number of people in the town?

4. To find out how many largemouth bass are in Dryden Lake in central New York State, a naturalist captured 213 largemouth bass and made a mark on their fins. The fish were returned to the lake. About a month later, the naturalist caught 104 bass, and 13 of them had marks (Chatterjee in Mosteller et al., *Statistics by Example: Finding Models*). Find the 95% confidence interval for the number of largemouth bass in the lake.

5. In the taxicab experiment in Application 27, observers noted 48 taxicabs. The next day, they saw 52 taxicabs, and 10 were those they had seen the day before.

   a. Find the 95% confidence interval for the number of taxicabs in the city.

   b. The true number of cabs was 420. Is this number in the confidence interval?

IX. THE GERMAN TANK PROBLEM

In the early years of World War II, American and British intelligence information about Germany's war production proved to be inaccurate and contradictory. Thus, in 1943, statisticians at the United States Embassy in London began trying to estimate German war production by analyzing the serial numbers on captured German equipment. On some types of equipment, such as tire molds and tank gearboxes, the Germans numbered the items sequentially 1, 2, 3, 4, and so on.

Suppose the Allies captured four German tanks that bore the serial numbers 41, 23, 43, and 52. What is the best estimate for the total number of tanks?

This problem is different from the one we have been doing throughout most of this book: estimating the population percentage. In that problem, we took a random sample and used the sample proportion to estimate the population percentage. The estimator to use, the sample proportion, was obvious, so we worked on how to find the sampling error. In the capture-recapture problem of Section VIII, we wanted to estimate the total population size, just as we do here. In capture-recapture, the *estimator* we used was

$$N = \frac{\left(\begin{array}{c}\text{number of marked fish} \\ \text{in the population}\end{array}\right)\left(\begin{array}{c}\text{number of fish} \\ \text{in the sample}\end{array}\right)}{\left(\begin{array}{c}\text{number of marked fish} \\ \text{in the sample}\end{array}\right)}.$$

However, the information we have available now is completely different.

In the German tank problem, the challenge is to choose a good estimator for the total number of tanks. It is not obvious how to construct an estimator. We must first think of several ways to estimate the number of tanks and then decide which estimator works best. We will again use simulation to help us solve this new problem.

Solving the German Tank Problem

Let's do some experiments to simulate the German tank problem. Your teacher has a container of objects numbered 1, 2, 3, and so on, up to N. Your job is to estimate the total number of objects, N. Without looking into the container, one student should capture a sample of three "tanks."

1. What are the numbers of the three tanks?

2. Write a *method* or a formula for estimating the total number of tanks in the container. (You may want to work with several other students.) This method or formula is your *estimator*.

3. Using your method, how many tanks do you estimate are in the container?

We are now going to see which group has the best method of estimating the total number of tanks.

4. Your teacher will write a chart like this one on the board. Copy it onto your paper.

Estimators

Trial						
1						
2						
3						
4						
5						

In the boxes at the top of the chart, write the methods suggested by the different groups in your class. Across from "trial 1," write the estimate of the number of tanks given by each method.

5. Replace the three tanks, mix the objects in the container, and have another student capture three tanks. What are the numbers of these three tanks?

6. Estimate the number of tanks in the container using your method. Pretend that you did not see the results from trial 1.

7. Place your estimate and those of the other groups in the row headed by "trial 2."

8. Repeat questions 5, 6, and 7 for trials 3, 4, and 5.

9. Look in the container. How many tanks are in it?

To determine which method is the best estimator, statisticians sometimes use a rule called *least squared error*. For example, suppose the container actually had 40 tanks in it, and your method produced an estimate of 42 on trial 1, 35 on trial 2, 40 on trial 3, 51 on trial 4, and 31 on trial 5. Your estimate was 2 too big on trial 1, 5 too small on trial 2, just right on trial 3, 11 too big on trial 4, and 9 too small on trial 5. Your errors are 2, −5, 0, 11, and −9. The total of your errors is −1. So that negative errors and positive errors don't cancel each other out like this, statisticians add up the squared values of the errors. In this example, the sum of the squared errors is 4 + 25 + 0 + 121 + 81, or 231.

10. Find the sum of the squared errors for each method in your table.

11. Which method has the smallest sum? Congratulate the students who invented this method.

Statisticians also like to use estimators that are *unbiased*—that is, estimators that do not consistently give answers that are too large (or too small).

12. Were any methods your class used biased? Try to determine why these methods were biased.

Now read the note at the end of this application to discover the method statisticians used during the war. This method worked out very well. The records of the Speer Ministry, which was in charge of Germany's war production, were recovered after the war. The table below gives the actual tank production for three different months, the estimate from serial number analysis, and the number obtained by traditional American/British intelligence gathering.

Month	Actual Number of Tanks Produced	Serial Number Estimate	Estimate by Intelligence Agencies
June 1940	122	169	1000
June 1941	271	244	1550
September 1942	342	327	1550

Source: *Journal of the American Statistical Association*, 1947.

13. During World War II, Allied statisticians also conducted a serial number study of tires on several German Mark V tanks to determine the production of one tire manufacturer. Each tire was stamped with the number of the mold in which it was made; captured tires had 20 different mold numbers from this manufacturer. The largest mold number was 77.

   a. What is the best estimate of the total number of molds?

   b. What additional piece of information do you need to estimate this manufacturer's daily tire production?

14. Suppose you are standing on a corner watching taxis go by. You see that the numbers of the first five taxis are 284, 570, 321, 319, and 35. What is the best estimate for the total number of taxis? List the assumptions you are making to get this estimate.

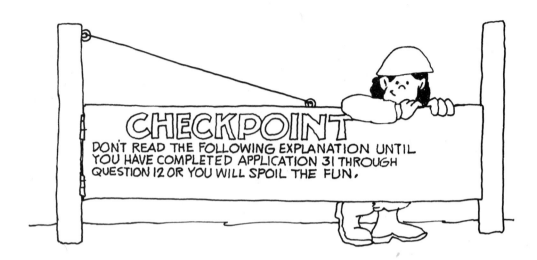

A good method for estimating the number of tanks is to multiply the largest tank serial number by $(n+1)/n$, where n is the number of tanks in the sample. For example, if the Allies captured four tanks with serial numbers 41, 23, 43, and 52, the estimate for the total number of tanks would be

$$\frac{5}{4}(52) = 65.$$

Among all estimators that are unbiased, this estimator is the best because it tends to minimize the sum of the squared errors.

The explanation of why this estimator is reasonable is simple. Suppose, for example, that we capture four tanks. Imagine each tank on a number line above its serial number:

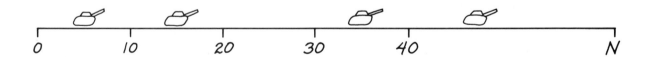

These four tanks divide all of the tanks into five groups. We would expect that the serial number, m, of the largest-numbered tank captured would be 4/5ths of that of the last tank in the population, N. That is,

$$m \cong \frac{4}{5}N$$

so
$$N = \frac{5}{4} m \ ,$$

which is our formula

$$N = \left(\frac{n+1}{n} \right) m$$

where N is the total number of tanks,
n is the number of tanks captured,
m is the largest serial number of the captured tanks.

Alternatively, we can write the formula

$$N = \left(\frac{n+1}{n} \right) m$$

$$= \frac{nm + m}{n}$$

$$= m + \frac{m}{n}$$

which can be interpreted as adding the average gap between serial numbers to the largest serial number.

Finding the Confidence Interval for the German Tank Problem

So far we have learned to estimate the total number of German tanks by multiplying the maximum of the n observed serial numbers by $(n+1)/n$. It would be useful to construct a confidence interval for the number of tanks, just as we formed confidence intervals for the sample survey and capture-recapture problems. This application shows how we can use the same methods we used for the other two problems to get a confidence interval here.

Let's call the unknown number of tanks (the size of the population) N. We get the maximum of the n observed serial numbers from the sample; call this maximum m. We know both n and m but not N.

The steps that follow are exactly the same as those in the sample survey problem. First we simulate the sampling distribution of the observation m, assuming some population size N and sample size n. Then we summarize the sampling distribution by a 90% box plot. Next we arrange the 90% box plots, for different values of N, in a chart. From these box plots we can read off, for each N, which values of m are likely sample maximums and which are unlikely. Moreover, by reading in the vertical direction, we can learn which values of N make the observed m a likely sample maximum. These values of N give the confidence interval.

We will obtain the confidence interval for question 13 of Application 31, in which a sample of size $n = 20$ gave a maximum value of $m = 77$. Clearly, N must be greater than or equal to 77.

First we estimate the sampling distribution of m for samples of size 20 for several different N. Let's start with $N = 77$. Using the random number table, we obtain a sample of 20 different values from the population 1, 2, 3, . . . , 76, 77. We need 20 different values because we can't catch the same tank twice. Here are the 20 values:

63, 28, 32, 15, 40, 61, 59, 01, 73, 33, 02, 50, 05, 12, 58, 49, 67, 42, 09, 51.

The maximum, m, is 73. We need the maximum in many samples, each of size 20, to get the sampling distribution. We did 40 trials and obtained these maximums:

73, 71, 76, 75, 69, 70, 76, 72, 77, 76, 76, 68, 74, 76, 71, 55, 76, 77, 67, 74, 76,

77, 56, 73, 73, 76, 75, 77, 68, 76, 75, 77, 68, 69, 61, 70, 77, 77, 75, 75.

The next step is to summarize the 40 trials for $N = 77$ by a 90% box plot. Using the 40 values listed above, and constructing the box exactly as we did in Section III, we get the following plot.

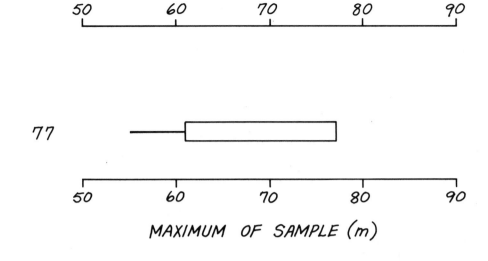

1. We must calculate the sampling distribution for many different population sizes N. We did this for $N = 77$. Now you will do it for $N = 85$. For $N = 85$, use the random number table to generate a sample of 20. What is the maximum, m?

2. Combine results from students in the class until you have generated 40 such samples for $N = 85$, and list the 40 sample maximums.

3. Take your 40 trials for $N = 85$ and construct the 90% box plot. Place this box plot above the one for $N = 77$ on a chart like the one above.

4. Why doesn't your box for $N = 85$ extend to the right of $m = 85$?

5. Is $m = 72$ a likely sample maximum for a population size of $N = 77$? For $N = 85$?

6.   a. Find a value of m that is a likely sample maximum for $N = 77$ but not for $N = 85$.

     b. Find a value of m that is a likely sample maximum for $N = 85$ but not for $N = 77$.

     c. Find a value that is likely for both $N = 77$ and $N = 85$.

Now you see what we must do. We have to fill in the chart from question 3 for many more values of N. But generating all these samples is boring. This is a good job for a computer! We used a computer to produce the following box plots. We could have given more values for N, but this chart will give you the idea.

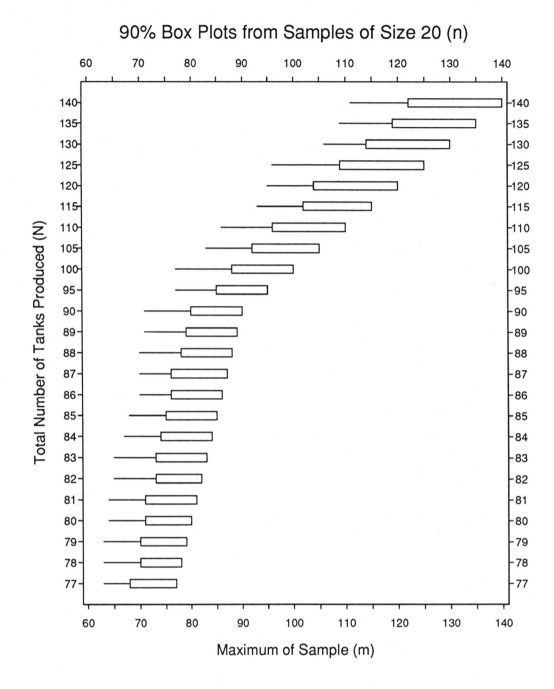

90% Box Plots from Samples of Size 20 (n)

Total Number of Tanks Produced (N)

Maximum of Sample (m)

Use this chart to answer the following questions.

7. a. Is $m = 80$ a likely sample maximum if $N = 90$?

   b. Is $m = 80$ a likely sample maximum if $N = 100$?

   c. List all the population sizes N that have $m = 80$ as a likely sample maximum.

To find the confidence interval for the total population size, we read down the chart as before. The 90% confidence interval includes all those populations that have the sample maximum m as a likely sample maximum.

8. What is the 90% confidence interval for the total population size if $m = 80$?

9. Give the 90% confidence interval for the total population size if $m = 85$.

10. From this chart, you cannot calculate the confidence interval for the total population size if $m = 125$. Explain why not.

11. Give the confidence interval for the total population size if $m = 77$, which applies to question 13 of Application 31. What estimate of N did you give then? Is your estimate inside this interval? Is it at the center of this interval?

12. What is the probability that an interval calculated in this way will contain the true population size N?

X. CONCLUSION

We hope that you have enjoyed working through this book and that you will now be able to understand and evaluate the surveys and polls you read about. For example, consider this article:

> "The Gallup poll, commissioned by the teachers' union, found that 57% of those surveyed believed their local schools were having a hard time attracting good teachers. Most pointed to low pay as the root of the problem . . . . Gallup questioned 1,501 adults by telephone in late April and May. The standard margin of error was plus or minus 3 percentage points" (Newark *Star-Ledger*, July 2, 1985).

Here are the important points to keep in mind when reading this article.

1. In any survey that involves a sample and not the entire population, we do not expect the result to be exact. If Gallup took a census of the whole adult population, the percentage of people who agreed with the statement would probably not be exactly 57%.

2. However, the larger the size of the sample, the closer the sample proportion tends to be to the true percentage.

3. Gallup reports a sampling error of 3%. This statement means that the true population percentage is probably somewhere between 54% and 60%. We say "probably" because the true population percentage is within the given interval in 95 out of every 100 such surveys. However, in 5 out of every 100 surveys, the true population percentage will be outside the given interval. This is what we can expect; it does not necessarily mean that the pollster made some mistake in conducting or analyzing the poll.

4. When we say that the true population percentage is probably between 54% and 60%, we mean that from any of these populations—with percentages either 54%, or 55%, or 56%, or 57%, or 58%, or 59%, or 60%—a sample proportion of 0.57 is likely.

5. The random variability from sample to sample is only one of several sources of error. In addition, bias can result because people may refuse to respond; they may not tell the truth; the survey question may be poorly worded; the timing of the survey may be bad; or the interviewer may make a mistake. When such sources of bias are present, it is always difficult and sometimes impossible to estimate how far the sample proportion is from the population percentage.

We hope that this book has also helped you understand and appreciate some of the basic concepts and methods of *statistical inference*. We used these methods throughout most of the book to solve the sample survey problem, but we also used exactly the same ideas to solve two other problems, capture-recapture and German tank.

All three of these problems involve the same key ideas. We want to learn about a specific *population*. We obtain a *random sample* from the population and use the sample to *estimate* what we want to know about the population. But estimating is not enough. We must also evaluate how good our estimate is. Since we don't know the precise population, we can't just compare the estimate to the population. What we can do, however, is use *simulation* to learn how the value of the estimate varies from one random sample to another. By simulating many samples from a population, we get the *sampling distribution* of the estimate. We sample from many different possible populations in order to get the sampling distribution for each. (More advanced mathematics and statistics courses use mathematical probability formulas instead of simulation to find the sampling distribution, but the overall approach is exactly the same.) Finally, from the sampling distributions, we can calculate a *confidence interval* for the population.

This book has shown you the fundamentals of making a statistical inference about a population from a sample. You can use these methods to solve many other statistical problems as well.

Bibliography

Surveys

Ferber, R., Sheatsley, P., Turner, A., and Waksberg, J. *What Is a Survey?* Washington, D.C.: American Statistical Association (no date). This 25-page booklet is available free from the American Statistical Association, 806 Fifteenth Street, N.W., Washington, D.C. 20005. It contains sections on types of surveys, designing and conducting a survey, and using the results of a survey.

Freedman, D., Pisani, R., and Purves, R. *Statistics*. New York: W. W. Norton, 1978. See Part VI on sampling.

Hollander, M., and Proschan, F. *The Statistical Exorcist: Dispelling Statistics Anxiety*. New York: Marcel Dekker, 1984. This book contains chapters about sampling, the *Literary Digest* poll, using a table of random numbers, the German Tank Problem (called the Racing Car Problem), and capture-recapture.

Moore, D. *Statistics: Concepts and Controversies*, 2nd ed. New York: W. H. Freeman, 1985.

Tanur, J. M. (Ed.). *Statistics: A Guide to the Unknown*, 2nd ed. San Francisco: Holden-Day, 1978. Chapters include "Opinion Polling in a Democracy" by George Gallup, "How Accountants Save Money by Sampling," "How to Count Better: Using Statistics to Improve the Census," and "Information for the Nation from a Sample Survey."

Williams, B. *A Sampler on Sampling*. New York: John Wiley, 1978.

Capture-Recapture

Chatterjee, S. Estimating the Size of Wildlife Populations. In Mosteller, F., Kruskal, W. H., Pieters, R. S., Rising, G. R., Link, R. F. (Eds.). *Statistics by Example: Exploring Data*. Reading, MA: Addison-Wesley, 1973.

Chatterjee, S. Estimating Wildlife Populations by the Capture-Recapture Method. In Mosteller, F., et al. (Eds.). *Statistics by Example: Finding Models*. Reading, MA: Addison-Wesley, 1973.

Swift, J. H. Capture-Recapture Techniques as an Introduction to Statistical Inference. In Sharron, S. (Ed.). *Applications in School Mathematics, 1979 Yearbook*. Reston, VA: National Council of Teachers of Mathematics.

The German Tank Problem

Larsen, R. J., and Marx, M. L. *An Introduction to Mathematical Statistics and Its Applications.* Englewood Cliffs, NJ: Prentice Hall, 1981. This calculus-based college textbook has a good discussion of estimators. See pages 193–194, 201–203, 207–211, 218, 224–227, 247–248.

Noether, G. E. *Introduction to Statistics: A Nonparametric Approach,* 2nd ed. Boston: Houghton-Mifflin, 1976. See pages 2–11 and 30–35.

Ruggles, R., and Brodie, H. An Empirical Approach to Economic Intelligence in World War II. *Journal of the American Statistical Association, 42,* 1947, 72–91. This completely nontechnical article makes fascinating reading.

Vannman, K. How to Convince a Student that an Estimator Is a Random Variable. *Teaching Statistics, 5,* May 1983, 49–54.

Data Sheet for Application 13

Population A

```
XOOOXXXXXOOOXXOXXXXO
OXOOOXOXXOOOOXOOXOOO
OXOOOXOXXOXXOXXXXXOX
OXXOOXOXXOOXOXOOOOOO
OXOOXXXOOOOOOOOOOOXX
OOXXOOOXXOOOOOOOXXXO
XXOOXOOXXXXOOOXXOOOO
XOXOOOOOOOOXOOOOOOO
XXOXXOXOOXOOOOXXOOXO
XXOOOXXOOXOXOOOOOOO
```

Population B

```
OOOOOOOOOOOOOXOOOOOX
OOOOOOOOOOOOOOOOOXOX
OOOOOOOOOOOOXOOOOOOO
OOOOOOOOOOOXOOOOXOOO
OOOOOOOOOOOOOOOOOOOO
OOOOOOOOOOOOOOOOOOOO
OOOOOOOOOXOOOOOOOOO
OOOOOOOOOOOOOOOOOOOO
OOOOOOOOOOOOOOOOOOOO
OOOOOOOOOOOOOOOOOOOO
```

Population C

```
OOOXOXOXOXXXOXOOXOOX
OOXOXOOOOOOXOOOOXOO
XXXXOOOXOOOOOXOOXXXO
OOOOXXOOXOOOXXOXOXXX
OXOOOOXOOOXXXXXOOXOO
XXOXOOXOOXXOOOOOXOXX
XXOOOOOOXOOXXXOOXXO
OXOOXOXOOOOXOOOOXOOO
XOOOXOOXXOOOXXOXOXOO
XOOXOOOOXOOXOOXXXOOO
```

Population D

```
XXOOOXXXOXXOXXOXXOXX
OXOXOXXXXOOOXXXXXXOO
XOXXXOXOOOOXOOXXOOOO
OXXXXXXXXXXXXXXXXXX
XXOOXXXOXOXOXXOXXXXO
OOOXOOXOXXXXXXXXOXXX
XXOOXXXOXXOXOXOOXXOX
OXOOOOOOXXOXXXXOXOXO
OOOOOXOXOOOOXOOOXOXO
OOOOOOXOXOOOOOOXXOOX
```

Population E

```
XXOXXXXXXXOXXOXXXXXX
XXOXOXXXOXXXOXXXOOXX
XXOXXXOXOOXOOXXXOXX
XXOXXOXOXXXXXXOXXXX
XXOOXOXXOXXXXOXXXOX
XOXXXOXOXXXOXXXXXOX
OXOOXOXXXXXOXOXOOXX
OXXOXXOOOXOOOXXXXOX
XXXOXXOXXOXXXXXXXOX
XOOXXOOXXXXXOXXXXXX
```

Population F

```
XXXXOXXXXXXXXXXXOOOOX
XXXXXXXXXXXXXXXXXXX
XXXXOXXXXXXXXXXXXOOXX
XOOXXXXXXXXXXOXXXXXOX
XOXXXXOXXXXXXXXXXXXX
XXXXXXXOXXOXXXXXXXX
OXXXXOXXXXXXXXOXXXOXX
XXXXXXXOXXXXXOXXXXXX
XXXXXOXXXXXOXXXXXXXX
XOXOXXXXXOOXXXXXXXXX
```

Population G

```
XXXXXXXXXXXXXXXXXXXX
XXXXXXXXXXXXXXXXXXXX
XXXXXXXXXXXXXXXXXXXX
XXXXXXXXXXXXXXXXXXXX
XXXXXXXXXXXXXXXXXXXX
XXXXXXXXXXXXXXXXXXXX
XXXXXXXXXXXXXXXXXXXX
XXXXXXXXXXXXXXXXXXXX
XXXXXXXXXXXXXXXXXXXX
XXXXXXXXXXXXXXXXXXXX
```

Population H

```
OOOOOXOOOOOOOXOOOOO
OOOXXOXXXOOOOOOXOOOX
OXOOOOOOOOOOXXOXOOO
OXOXOOXOXOOOOOOOXOOO
OOOXOXOOOOOOOXOOOOOO
OOOOOXOXXOOOOOOOXOOO
OOOOOOOOOXOOOOOOOOX
OOOXOOOOOOOXOOOOOOO
OOOOXOOXOOOOOOOOOOO
OXOOOOOOOXOOOOOOOOX
```

Population I

```
OOOOOOOOOOOOOOOOOOOO
XXOOOOOOXOOOOOOOOOOX
OOOOOOOOOOOOOOOOOOOX
OOOXOOOOOXOOOOOXOOXO
OOOOOOOOOOOOOOOXOOO
OOOOOOOOOOOXOOOXOOO
OXOOOOOOOOOOOOOOXOO
OXOOOOOOOOOOOOOOXOO
OOOOOOOOOOOOOOOOOOO
OOOOOOOOOXOOOXOOOOOX
```

Population J

```
OOXOOOOXOXOXXXXOXOXX
OOXXOOXOOXOOOOXXXXX
XXOOXOOXOXOXXXOXXOO
OOXOXXXXOOXOOOOXXXOX
OXOXOOOXOXOOXOOXXOXO
XXOXOOXXXOOOXOOOXOOX
OXOOXOXOXXXOXOOXXXOX
XOOOOOXXOXXXXXOOXOO
XOOOOXOXXOOXOOOXXXO
XXOXXOXXOXOXXXOXOXOX
```

Population K

```
XXXXXXXXXXOOXOXXXXXO
XXOXOXOXOOXXXXXXXOXX
XXXOXXXXXOXXXXXXOXXO
XXXXXOXXXXXXXOXOOXXX
XXXXXOOXOOOOXXXXXXX
OXOXXXXXOXXXXXXXOOO
XXOXXXXXXOOXOXOXXOX
XOXXOXOXXOXXXXXXOOXX
OXXOOXXOXXXXXXOXXOOX
XXXXXXOOXXXOXXOOXXXX
```

Population L

```
OOOOOXXOOXOOOXOXXOOO
OOOXXOXXOOOOXOXXOXOO
OOOXXOXOOOXOOOOOXOOX
OOOOOOOOOXXOXXOOOOXO
XOOOOOOOOOXOOOOXOOXO
OXOOXOOOOOXXOOOOXOXO
OOXXOOXOOOOXOOOOXXOO
XOXOOOXOOXXOXOOOOOOX
OXXOOXXOOOOXOOOOOOXO
OOOOOXOXOOOOOOOOXOOO
```

Table of Random Numbers

39634	62349	74088	65564	16379	19713	39153	69459	17986	24537
14595	35050	40469	27478	44526	67331	93365	54526	22356	93208
30734	71571	83722	79712	25775	65178	07763	82928	31131	30196
64628	89126	91254	24090	25752	03091	39411	73146	06089	15630
42831	95113	43511	42082	15140	34733	68076	18292	69486	80468
80583	70361	41047	26792	78466	03395	17635	09697	82447	31405
00209	90404	99457	72570	42194	49043	24330	14939	09865	45906
05409	20830	01911	60767	55248	79253	12317	84120	77772	50103
95836	22530	91785	80210	34361	52228	33869	94332	83868	61672
65358	70469	87149	89509	72176	18103	55169	79954	72002	20582
72249	04037	36192	40221	14918	53437	60571	40995	55006	10694
41692	40581	93050	48734	34652	41577	04631	49184	39295	81776
61885	50796	96822	82002	07973	52925	75467	86013	98072	91942
48917	48129	48624	48248	91465	54898	61220	18721	67387	66575
88378	84299	12193	03785	49314	39761	99132	28775	45276	91816
77800	25734	09801	92087	02955	12872	89848	48579	06028	13827
24028	03405	01178	06316	81916	40170	53665	87202	88638	47121
86558	84750	43994	01760	96205	27937	45416	71964	52261	30781
78545	49201	05329	14182	10971	90472	44682	39304	19819	55799
14969	64623	82780	35686	30941	14622	04126	25498	95452	63937
58697	31973	06303	94202	62287	56164	79157	98375	24558	99241
38449	46438	91579	01907	72146	05764	22400	94490	49833	09258
62134	87244	73348	80114	78490	64735	31010	66975	28652	36166
72749	13347	65030	26128	49067	27904	49953	74674	94617	13317
81638	36566	42709	33717	59943	12027	46547	61303	46699	76243
46574	79670	10342	89543	75030	23428	29541	32501	89422	87474
11873	57196	32209	67663	07990	12288	59245	83638	23642	61715
13862	72778	09949	23096	01791	19472	14634	31690	36602	62943
08312	27886	82321	28666	72998	22514	51054	22940	31842	54245
11071	44430	94664	91294	35163	05494	32882	23904	41340	61185
82509	11842	86963	50307	07510	32545	90717	46856	86079	13769
07426	67341	80314	58910	93948	85738	69444	09370	58194	28207
57696	25592	91221	95386	15857	84645	89659	80535	93233	82798
08074	89810	48521	90740	02687	83117	74920	25954	99629	78978
20128	53721	01518	40699	20849	04710	38989	91322	56057	58573
00190	27157	83208	79446	92987	61357	38752	55424	94518	45205
23798	55425	32454	34611	39605	39981	74691	40836	30812	38563
85306	57995	68222	39055	43890	36956	84861	63624	04961	55439
99719	36036	74274	53901	34643	06157	89500	57514	93977	42403
95970	81452	48873	00784	58347	40269	11880	43395	28249	38743
56651	91460	92462	98566	72062	18556	55052	47614	80044	60015
71499	80220	35750	67337	47556	55272	55249	79100	34014	17037
66660	78443	47545	70736	65419	77489	70831	73237	14970	23129
35483	84563	79956	88618	54619	24853	59783	47537	88822	47227
09262	25041	57862	19203	86103	02800	23198	70639	43757	52064

Table of Random Numbers

59718	77768	50032	53440	41359	33021	01938	86092	87426	80010
91977	35682	34043	26290	40447	12411	32837	12151	21227	81491
88224	92826	92683	66928	95518	70106	92397	62132	97206	26324
01288	56565	78378	72344	12566	58325	40257	93212	49208	51320
19483	45024	12857	46267	94007	98674	54199	29738	24084	91964
33652	12588	55326	05702	43815	61284	13606	65461	70415	91440
32207	57357	18841	61415	57755	46846	41422	35285	37870	55929
99945	87321	41676	70537	39314	45154	93823	14053	81888	11464
29773	64388	95180	80750	12815	77661	89578	42194	99329	21247
92329	55414	05162	94197	19267	68846	27895	12005	80292	49745
75834	71767	45378	40316	61259	13140	66115	61564	76757	62599
22755	89933	41019	18996	13005	31853	72795	22193	59897	62049
09056	73260	95209	33157	15608	37565	93590	85486	80932	76059
66250	96883	74585	74550	89984	28356	77938	69704	19034	19744
37052	83115	38995	52825	93308	75276	21274	48777	75400	62004
81653	74197	85789	50614	52742	48213	94759	80701	08234	44686
41417	37426	42282	34323	83341	38345	83018	25015	68282	94820
27862	25188	15227	90981	06296	86815	04322	44750	01554	91302
85083	13672	29208	17587	12217	24032	52318	83860	81936	29114
05649	48381	63320	11822	11590	75112	54027	56579	81397	14691
91654	28637	01627	24482	33119	29924	69390	85040	66927	63521
43540	82299	18928	35588	55113	78385	61536	49596	05202	40993
33276	99974	62800	97999	56683	61505	85617	32656	16834	88980
18139	96834	07488	32049	53532	12159	75508	10924	25298	96474
07403	42795	55422	49346	44612	61632	81241	04660	95163	16285
05374	34289	66087	74636	64247	73598	42730	79472	79834	72702
63121	17926	84377	16927	91950	26475	10086	61879	03475	64750
66148	59081	34743	69023	50306	63739	14717	32374	19119	96284
92153	23320	34180	78025	42391	35908	73996	49173	47360	92856
06629	93991	80847	49133	45105	34818	10122	31369	33312	94856
74784	07080	13104	64110	98440	56468	88959	67988	58764	70414
59043	74797	24791	65130	97918	99820	32673	44512	36847	14028
58572	79127	74870	47218	03752	92434	71791	28040	60536	37429
75069	76687	43795	50161	20794	95015	42376	33178	10265	03394
72258	09820	54814	84454	32761	59316	14974	80017	37524	25760
16186	64983	27652	53966	75826	16790	13767	52267	65505	56954
54047	17961	92967	27968	12463	85270	13763	96297	43279	93087
42301	36874	19357	14982	22806	69213	79929	48973	21969	28172
87940	43389	26009	52702	03148	70789	88539	19084	59200	88168
91551	24267	81423	17461	09300	11928	98793	97748	95430	11644
03166	69589	65596	56997	70092	63418	92825	91586	76847	51167
64280	45356	96248	79274	15733	72317	44107	80124	99672	44523
28464	37825	88800	20180	28989	75914	46882	28736	60408	63180
36861	76806	80789	30886	71013	56044	52405	81063	04283	41256
43125	34876	18177	22382	37920	77067	93319	29881	37050	32533

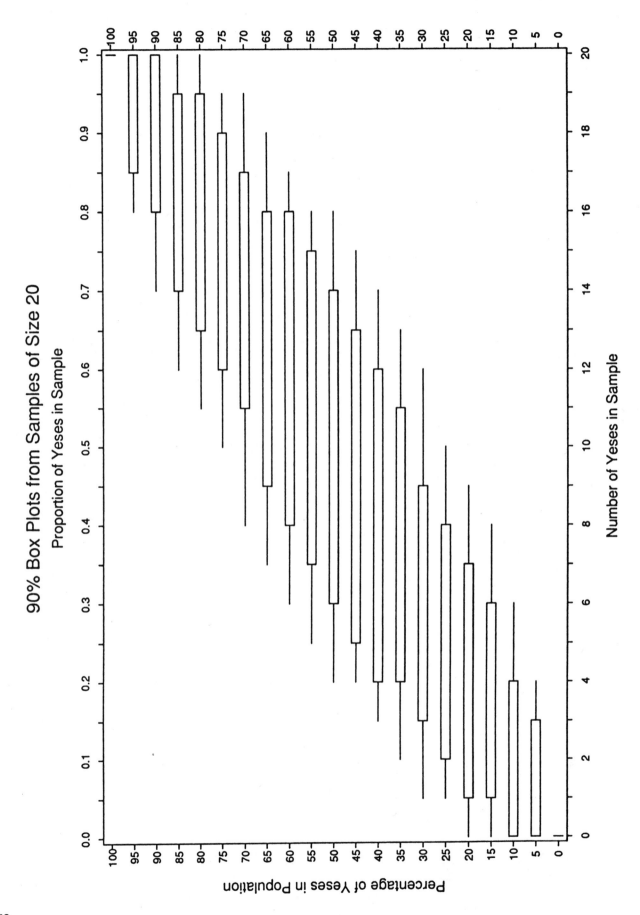

90% Box Plots from Samples of Size 20
Proportion of Yeses in Sample

Percentage of Yeses in Population

Number of Yeses in Sample

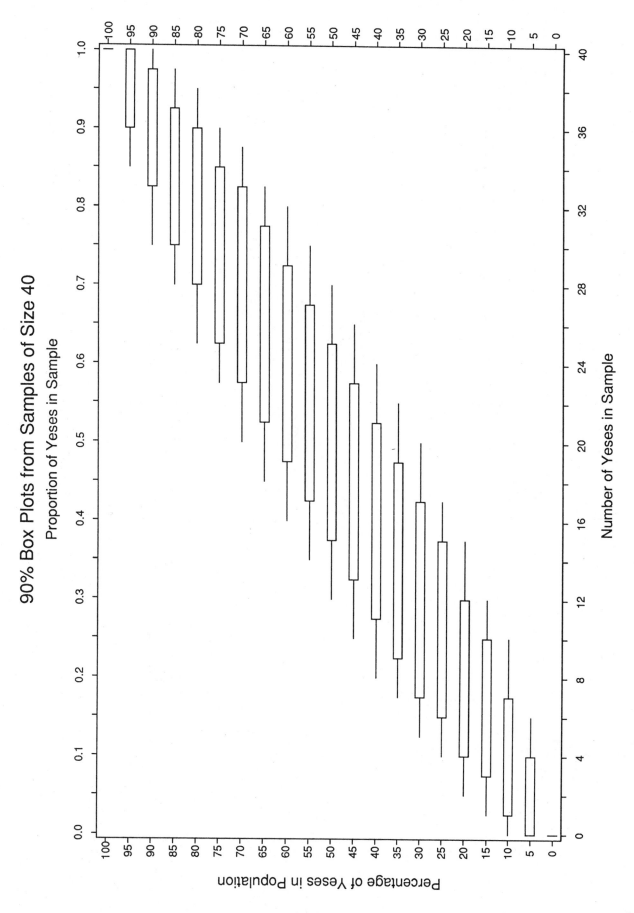

90% Box Plots from Samples of Size 40
Proportion of Yeses in Sample

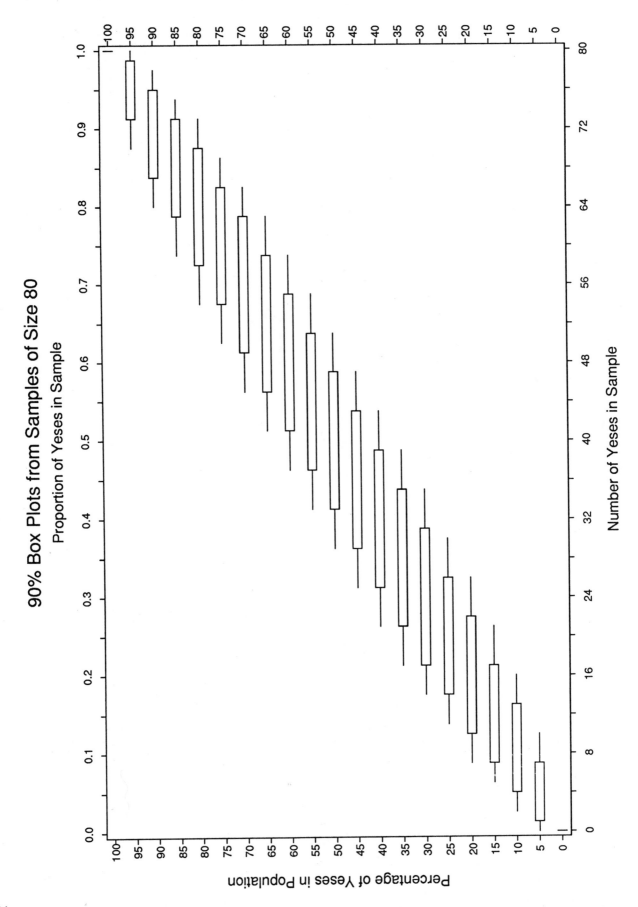

90% Box Plots from Samples of Size 80
Proportion of Yeses in Sample

Percentage of Yeses in Population

Number of Yeses in Sample

90% Box Plots from Samples of Size 100
Proportion of Yeses in Sample

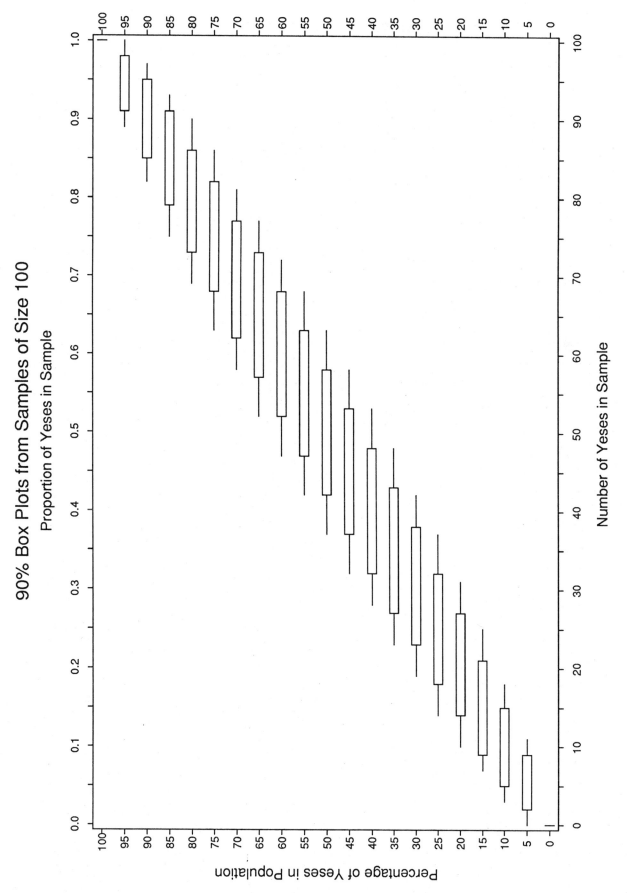

Percentage of Yeses in Population

Number of Yeses in Sample

INDEX OF STATISTICAL TERMS